Reasonable Books = Reasonable Prices

Irina Kleyman

Exceptional for Home Schooling

Math Help
Grade 8

AF813128

First Edition

Reasonable Books = Reasonable Prices

Irina Kleyman

Exceptional for Home Schooling

Math Help Grade 8

I dedicate this book to my children and grandchildren.

Acknowledgments
Special thanks to Michael G LaFratta, Vladimir Kleyman, who has assisted in bringing all 10 books to market.

Author's note: In the many years that I've taught mathematics, I have been informed by many of my students that my methods for teaching have helped them tremendously in learning math. I have written these books in order to continue to help others learn math effectively.

Math Help Grade 1
Math Help Grade 2
Math Help Grade 3
Math Help Grade 4
Math Help Grade 5
Math Help Grade 6
Math Help Grade 7
Math Help Grade 8
Math Help for College (Summary Grades 1 – 5)
Math Help for University (Basic, Algebra, Geometry)

ISBN: 978-1-7349699-7-9 Paperback

Syllabus Math 5, 6, 7, 8

Grade 5 # 1–5, Grade 6 # 6–10, Grade 7 # 11–15, Grade 8 # 16–21

	Grades	Topics, Advanced Problems (A-2), Pages (p)
#	Chapters	Basic Problems (B), Chapter (C), Lessons (L)
1	5-1	Four–digit numbers (+ or -) (B)
2	5-2	Multiplying numbers (B)
3	5-3	Dividing numbers (B)
4	5-4	Four operations with numbers (B)
5	5-5	Finding unknown (A-5)
	Grade 5	Final Exam 5 B, C 5, L 16 #1-10 all, p 150
6	6-1	Many-digit numbers (+ or -) (B)
7	6-2	Multiplying numbers (B)
8	6-3	Dividing numbers (B)
9	6-4	Four operations with numbers (B)
10	6-5	Finding Unknown (A-6)
	Grade 6	Final Exam 6 B, C 5, L 16 # 1-10 all, p196
11	7-1	Whole numbers (B)
12	7-2	Properties of whole numbers (B)
13	7-3	Decimals (B)
14	7-4	Fractions and mixed numbers (B)
15	7-5	Converting fractions and decimals (B)
	Grade 7	Final Exam 7B, C 5, L12 #1-10 all, p. 209-210
16	8-1	Rational numbers (B)
17	8-2	Equations (B)
18	8-3	Geometry (B)
19	8-4	Word Problems (B)
20	8-5	Summary of Basic Math (B)
21	8-6	Extra Review (B)
	Grade 8	Final Exam 8 B, C 5, L14 #1-10 all, p.281-282

Syllabus includes topics, what students need to know.

Middle School can start from 5th, 6th, or 7th grade. Do not worry about school. Choose your grade, work hard, you have knowledge.

Grade 8
Chapter 1

Rational Numbers

www.mathbooksbyirina.com

Parents, Tutors, Students Middle Schools, High Schools, Colleges, University can help primary school students. Please read slowly math books to primary school students.

Grade 8 Chapter 1 Contents
Rational Numbers
Lesson 1 – 10
Basic Problems

Work with lesson 1, read rules and examples, 2) Solve the problems from contents, 3) Check answers in the end of chapter 1, 4) Work with the next lessons 2, 3, 4, 5, …, 5) Solve Test 1 A, 6) Grading Scale: A =10 points, B=9points, C=8 points, D=7 points, F=0-6 points, 7) If you solve 10 problems correctly, you get an A, 8) If have an A, you can move on to next chapter, 9) If your grade is lower than an A, work with the mistakes, then solve Test 1B, 10) When you finish all of the chapters, solve Final Exam in last chapter.

Lesson 1 Memorize table perfect square numbers. Solve the Pretest 1 # 1 – # 5
Lesson 2 Addition and Subtraction of Rational numbers
Solve Review 1 A # 16 - # 20, Solve Review 1 B # 16 – # 20

Lesson 3 Multiplication, Division and Powers of Rational numbers
Solve Review 1 A # 16 - # 20, Solve Review 1 B # 16 - # 20
Lesson 4 Order of five operations
Solve Review 1 A # 16 – # 20, Solve Review 1 B # 16 – # 20
Solve Review 2 A # 16 – # 20, Solve Review 2 B # 26 – # 30

Lesson 5 Laws for rational numbers
Solve Review 1 A # 26 – # 30, Solve Review 1 B # 26 – # 30
Lesson 6 Table 7 – 8. Solve Mental Math 9, 10, 11, 12 odd

Lesson 7 Square Roots.
Solve Review 1 A # 17 – # 20, Solve Review 1 B # 37 – # 40
Solve Review 1 C # 17 - # 20, Solve Review 1 D # 37 - # 40
Mental Math 13, 14 odd
Lesson 8 Worksheets 4
Solve Worksheet 4 A # 1 – # 10 all, Solve Worksheet 4 B # 1 – # 10 all

Lesson 9 Solve Review 1 A # 1 - # 10 all Solve Review 1 B # 1 - # 10 all
Lesson 10 Solve Test 1 A # 1 – # 10 all Solve Test 1 B # 1 – # 10 all
Answers Chapter 1

Grade 8 Chapter 1 Lesson 1

Table 2

Perfect Square Numbers

Take number 20 from the left column add number 5 from the first row to get number 25

Answer is in intersection row of 20 and column of 5

$25^2 = 25 \cdot 25 = 625$

Memorize Table of Perfect Square for Indicated Numbers

a^2	0	1	2	3	4	<u>5</u>	6	7	8	9
10	<u>100</u>	<u>121</u>	<u>144</u>	<u>169</u>	<u>196</u>	<u>225</u>	<u>256</u>	<u>289</u>	<u>324</u>	<u>361</u>
<u>20</u>	<u>400</u>	<u>441</u>	<u>484</u>	<u>529</u>	<u>576</u>	<u>625</u>	<u>676</u>	<u>729</u>	<u>784</u>	<u>841</u>
30	<u>900</u>	961	1024	1089	1156	1225	1296	1369	1444	1521
40	<u>1600</u>	1681	1764	1849	1936	2025	2116	2209	2304	2401
50	<u>2500</u>	2601	2704	2809	2916	3025	3136	3249	3364	3481
60	<u>3600</u>	3721	3844	3969	4096	4225	4356	4489	4624	4761
70	<u>4900</u>	5041	5184	5329	5476	5625	5776	5929	6084	6241
80	<u>6400</u>	6561	6724	6889	7056	7225	7396	7569	7744	7921
90	<u>8100</u>	8281	8464	8649	8836	9025	9216	9409	9604	9801

Grade 8 Chapter 1 Lesson 1
Pretest
Evaluate

1) $-3\frac{1}{3} \times 1\frac{1}{5} - 1\frac{1}{5} \times 1\frac{2}{3}$

1) ________

2) $1\frac{1}{4} \div (-4\frac{5}{8} + 3\frac{1}{16})$

2) ________

3) $(-2\frac{1}{5} + 0.6) \div 3.2$

3) ________

4) $-\frac{4}{25} \div (-0.6 + \frac{4}{5})$

4) ________

5) $-\frac{9}{20} \div (-0.3 + \frac{7}{20})$

5) ________

How do you feel about Math?

a) Love b) Like c) Nothing d) Do not Like e) Hate

Grade 8 Chapter 1 Lesson 2
Addition and Subtraction Rational Numbers

A line with a chosen origin, unit and direction called a number line. Origin is point O represents the number zero.
Numbers standing to the right of zero called positive numbers.
Positive numbers written with a plus sign in front of the numbers or written without a sign at all.

Numbers standing to the left of zero called negative numbers.
Negative numbers must always have a minus sign in front of them.
The number zero is not a positive or a negative number.
A number line is a line where any points on it represent numbers.

A number that corresponds to a given point on the number line called a coordinate of point. A (-2) O (0) B (4) - 2 0 4

An example of a number line is a thermometer for measuring temperature.

Opposite Numbers

Two numbers, differing from each other only in signs, called opposite numbers. Opposite numbers found at the same distance from the origin but in the opposite directions. For every number there is only one opposite number. The number zero is opposite into itself.
The opposite number of 2 is – 2 and the opposite number of - 2 is 2.
Minus is the word used for "opposite":

– (- 2) = 2 - (+ 2) = - 2 - (+ a) = - a - (- a) = a

Natural numbers, their opposites and zero called integers numbers
… -2, -1, 0, 1, 2 …
Each integer can show on a number line:

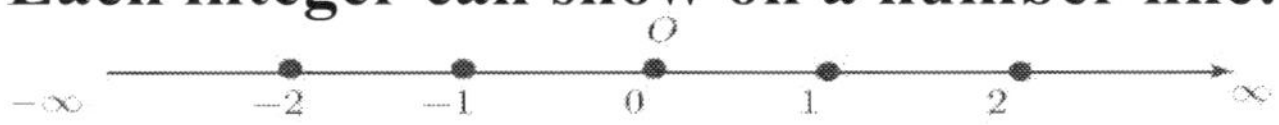

Integer's numbers, decimals, fractions together represent rational numbers. Rational numbers consist of all numbers that can expressed as a terminating decimal or as a repeating decimal. The irrational number can be in the form of an infinite non-periodic fraction. The real numbers include all the rational numbers, such as the integer −5 and the fraction 4/3, and all the irrational numbers, such as √2 (1.414..., the square root of 2).

Grade 8 Chapter 1 Lesson 2

Absolute Value

The distance from the origin to a point that represents a number called the absolute value of the number.
The absolute value of a number is always a positive number.

The absolute value of a positive number and zero is itself.
The absolute value of a negative number is opposite its number.
Opposite numbers have equal absolute value:

$|+4| = 4$ $\qquad$ $|0| = 0$ $\qquad$ $|-4| = 4$

Comparison of Numbers

In a number line, a point with a greater coordinate lies further to the right than a point with a small coordinate. A (-2) O (0) B (4)
Any positive number greater than zero: $4 > 0$ $\quad a > 0$
Any negative number less the zero: $-2 < 0$ $\quad a < 0$
Any positive number greater than any negative number: $4 > -2$
Any negative number less than any positive number: $-2 < 4$
To compare two negative numbers, use this rule: the smallest number has the greater absolute value: $-2 > -4$ **because** $2 < 4$
The order of numbers means putting numbers from smaller to larger number:
Example: - 4, - 2, 0, 2, 4

Addition and Subtraction

Two add two numbers with the same signs, put in the result, the sign of the addends and add the absolute value of the numbers:
$(+0.8) + (+1.25) = +2.05$ $\qquad$ $(-1.25) + (-0.8) = -1.25 - 0.8 = -2.05$
To add two numbers with different signs, put in the result, the sign of the addend with the greater absolute value and subtract its absolute value:
$0.8 - 1.25 = -0.45$ $\qquad$ $-1.25 + 0.8 = -0.45$
The sum of two opposite numbers equals zero: $-4 + 4 = 0$ $\quad -A + A = 0$
Subtracting numbers: $A - B = A + (-B)$, $-7-(-4) = x$ $\quad x+(-4)=-7$ $\quad x = -3$
To subtract two numbers, add to the first number the opposite of the second number: $-1.25 - (-0.8) = -1.25 + 0.8 = -0.45$
Rational numbers consist of all numbers that can expressed as a terminating decimal or as a repeating decimal.

Grade 8 Chapter 1 Lesson 2
Review 1 A

Add or Subtract of Rational Numbers
Add and Subtract in fractions but final answers put in decimals

1) $-\frac{5}{6}+(-\frac{1}{6})$

2) $-\frac{3}{4}-(-1\frac{1}{4})$

3) $-3\frac{1}{3}-(+5\frac{2}{3})$

4) $-2\frac{1}{6}-(+3\frac{5}{6})$

5) $-(+\frac{3}{8})-(+2\frac{5}{8})$

6) $-(-4\frac{3}{5})-(-3\frac{2}{5})$

7) $-\frac{3}{4}+(-1)$

8) $-\frac{3}{8}+(-\frac{1}{8})$

9) $-5\frac{3}{5}-(-4\frac{2}{5})$

10) $-\frac{5}{8}-(-2\frac{1}{8})$

11) $-1\frac{5}{8}-(-3\frac{1}{8})$

12) $-\frac{5}{8}-(-\frac{1}{8})$

13) $-3\frac{3}{8}+(-1\frac{1}{8})$

14) $-\frac{7}{8}-(-\frac{1}{8})$

15) $-1-(-\frac{2}{5})$

16) $-5\frac{3}{5}-(-9\frac{2}{5})$

17) $3\frac{2}{5}+(-8\frac{1}{5})$

18) $-4-(-3\frac{2}{5})$

19) $2\frac{3}{10}-(+4\frac{1}{10})$

20) $\frac{1}{5}+(-2)$

Grade 8 Chapter 1 Lesson 2
Review 1 B
Add or Subtract of Rational Numbers
Add and Subtract in fractions but final answers put in decimals

1) $-\frac{1}{4}-\frac{1}{8}$

2) $4\frac{3}{5}-(+2\frac{2}{10})$

3) $-1\frac{1}{5}-2\frac{3}{10}$

4) $2\frac{1}{3}-3\frac{5}{6}$

5) $-4+3\frac{2}{5}$

6) $\frac{3}{5}+(-\frac{7}{10})$

7) $3\frac{3}{8}-(+2\frac{3}{4})$

8) $-(+\frac{3}{8})+(-2\frac{3}{4})$

9) $-4\frac{1}{8}+3\frac{3}{4}$

10) $-1\frac{2}{3}-(+3\frac{5}{6})$

11) $-(-3\frac{1}{10})-(+2\frac{3}{5})$

12) $-\frac{5}{6}-(-1\frac{1}{3})$

13) $-2\frac{5}{6}+3\frac{1}{12}$

14) $-1\frac{5}{8}-(-3\frac{1}{4})$

15) $1\frac{1}{4}+(-2\frac{1}{8})$

16) $(-2\frac{2}{3})-(+\frac{1}{6})-(-2\frac{1}{3})$

17) $-5\frac{2}{5}-(-4\frac{2}{5})-(-3\frac{3}{5})$

18) $-\frac{1}{5}-(-\frac{3}{10})-(-2\frac{1}{10})$

19) $3\frac{1}{12}-(-2\frac{5}{6})-(-1\frac{1}{3})$

20) $-(+1\frac{3}{10})-(-2\frac{7}{10})-(-\frac{1}{5})$

Grade 8 Chapter 1 Lesson 3
Multiplication and Division Rational Numbers

Multiplication of Rational Numbers

Problem 1

The thermometer shows 0 F, the temperature of air decreases every hour by 2F. What was temperature of the air 3 hours ago? (-2)(-3)=6F

Problem 2

The thermometer shows 0 F, the temperature of air increases every hour by 2F. What was temperature of the air 3 hours ago? (+2)(-3)=-6F

To multiply two numbers with the same signs, the product is positive and multiplies the absolute values of the numbers:

	(+ 6)(+ 2) = + 12	(– 6) (- 2) = + 12
in short	(+) (+) = +	(-) (-) = +

To multiply two numbers with different signs, the product is negative and multiplies the absolute values of the numbers:

	(+ 6)(- 2) = - 12	(- 6) (+ 2) = - 12
In short	(+) (-) = -	(-) (+) = -

If the product of two numbers equals one, these numbers called reciprocals:

$(-2)(-\frac{1}{2})=1$ $\qquad$ $(a)(\frac{1}{a})=1$

The product of any number and one is the same number:

A · 1 = A $\qquad$ **(- 5) (1) = - 5**

The product of any number and zero is zero:

A · 0 = 0 $\qquad$ **(- 5)(0) = 0**

The product of zero and any number is zero:

0 · A = 0 $\qquad$ **0 (- 6) = 0**

Multiply decimals by 10, 100, 1000 …

(- 0.25) (10) = - 2.5 $\qquad$ **(0.25) (- 100) = - 25** $\qquad$ **(- 0.25) (- 1000) = 250**

Multiply decimals:

Put in first place number with more digits for easy multiplication.

(- 4) (2.5) = - 10 $\qquad$ **8 (- 12.5) = - 100** $\qquad$ **(1.25) (- 0.8) = - 1**

Grade 8 Chapter 1 Lesson 3

Division of Rational Numbers

To divide two numbers (-15) ÷ (-5) = x we need x (-5) = -15.
This means x = + 3 or (-15) ÷ (-5) =+3
To divide two numbers with the same signs, the quotient is positive and divides the absolute values of the numbers:

(+ 6) ÷ (+ 2) = + 3 (– 6) ÷ (- 2) = + 3

in short (+) ÷ (+) = + (-) ÷ (-) = +

To divide two numbers with different signs, the quotient is negative and divides the absolute value of the numbers:

(- 6) ÷ (+ 2) = - 3 (+ 6) ÷ (- 2) = - 3

in short (-) ÷ (+) = - (+) ÷ (-) = -

Any number divided by 1 is the same number:
A ÷ 1 = A (- 5) ÷ 1 = - 5
Any number other than zero divided by itself is 1:
A ÷ A = 1 (- 5) ÷ (- 5) = 1
Zero divided by any number other than zero is zero
0 ÷ A = 0 0 ÷ (- 5) = 0
Division by zero not allowed:
A ÷ 0 no solutions - 5 ÷ 0 none numbers 0 ÷ 0 undefined

Divide by 10, 100, 1000…
(- 5) ÷10 = - 0.5 5 ÷ (- 100) = - 0.05 (- 5) ÷ (- 1000) = 0.005
Divide the numbers:
(- 2) ÷ 5 = - 0.4 0.2 ÷ (- 5) = - 0.04 (- 2) ÷ (- 0.5) = 20÷5 = 4
0.2 ÷ (- 0.5) = (- 2) ÷ 5 = - 0.4 (- 1.2) ÷ 0.8 = (- 12)÷8 = - 1.5

Power of Numbers

Positive numbers in any exponent are always positive numbers:
$3^2 = 3 \cdot 3 = 9$ $2^3 = 2 \cdot 2 \cdot 2 = 8$
Negative numbers in an even exponent are always positive numbers:
$(-2)^4 = (-2)(-2)(-2)(-2) = 16$ $(-0.1)^2 = (-0.1)(-0.1) = 0.01$
Negative numbers in an odd exponent are always negative numbers:
$(-2)^3 = (-2)(-2)(-2) = -8$ $(-0.1)^3 = (-0.1)(-0.1)(-0.1) = -0.001$
An expression without parentheses: $-0.3^2 = -0.3 \cdot 0.3 = -0.09$
$-0.2^3 = -0.2(0.2)(0.2) = -0.008$ $-0.2^4 = -0.2(0.2)(0.2)(0.2) = -0.0016$

Grade 8 Chapter 1 Lesson 3
Review 1 A

Perform the indicated operations
Multiply in fractions but final answers put in decimals

1) $(+1\frac{2}{3})(+1\frac{1}{5})$

2) $(+\frac{3}{7})(-\frac{7}{12})$

3) $(-2\frac{4}{5})(-1\frac{3}{7})$

4) $2\frac{2}{5}(-1\frac{3}{12})$

5) $(-\frac{3}{5})(+1\frac{2}{3})$

6) $(-1\frac{5}{6})(1\frac{1}{11})$

7) $(+6\frac{1}{3})(-1\frac{2}{19})$

8) $(-1\frac{5}{6})(-\frac{3}{11})$

9) $(-1\frac{1}{7})(+1\frac{5}{16})$

10) $1\frac{1}{6}(-1\frac{1}{14})$

11) $-2.1(-\frac{5}{7})$

12) $-0.4(\frac{3}{5})$

13) $-3.2(1\frac{7}{8})$

14) $0.9(-1\frac{7}{9})$

15) $(-0.51)(-3\frac{1}{3})$

16) $\frac{5}{9}(-3.6)$

17) $-\frac{5}{6}(0.6)$

18) $-\frac{2}{3}(0.9)$

19) $(-2\frac{2}{7})(-1.4)$

20) $(-1\frac{2}{5})(-0.25)$

Grade 8 Chapter 1 Lesson 3
Review 1 B

Perform the indicated operations
Do operations in fractions but final answers put in decimals

1) $\frac{6}{7} \div (-\frac{3}{7})$

2) $(-8) \div \frac{4}{5}$

3) $3\frac{1}{3} \div (-\frac{5}{6})$

4) $(-2\frac{2}{3}) \div (3\frac{1}{3})$

5) $\frac{4}{5} \div (-2)$

6) $\frac{6}{7} \div (-2\frac{1}{7})$

7) $(-1\frac{2}{3}) \div (1\frac{1}{3})$

8) $(-3\frac{3}{5}) \div (-2\frac{2}{5})$

9) $(-1\frac{2}{7}) \div (-2\frac{1}{7})$

10) $(1\frac{2}{3}) \div (-1\frac{1}{3})$

11) $(-4.4) \div (2\frac{1}{5})$

12) $1.2 \div (-1\frac{1}{2})$

13) $(-3.2) \div (1\frac{1}{3})$

14) $(-4.5) \div (-1\frac{1}{2})$

15) $3.6 \div (-1\frac{11}{25})$

16) $-(-\frac{1}{5})^2 - (-\frac{2}{5})^2$

17) $(-\frac{1}{5})^2 - (-\frac{2}{5})^2$

18) $-(\frac{2}{5})^2 - (-\frac{3}{5})^2$

19) $(\frac{1}{4})^2 - (-\frac{3}{4})^2$

20) $(-\frac{3}{5})^2 + (-\frac{4}{5})^2$

Grade 8 Chapter 1 Lesson 4

Order of Five Operations

Operations have three levels: (+, - , ×, ÷, a^n)

Operations of the first level: addition (+) and subtraction (-)

Operations of the second level: multiplication (×) and division (÷)

Operations of the third level: power (a^n), root, and logarithm

Arithmetic expressions contain numbers, sign of operations, and sometimes parentheses.

Order of Operations with the Same Level of Operations

Expressions without parentheses that have operations of only addition or of only multiplication can done in any order:

-80 - 60 - 20 = - 80 – 20 - 60= - 160 -25(- 6) (- 4) = - 25×4×6= - 600

Expression without parentheses that have operations at only the same level of operations can done in order from left to right:

-80 - 40 + 20 = -120 + 20 = -100 -160 ÷ (- 4) × (- 2) =40× (- 2) = -80

Order of Operations of Two Levels of Operations (+, - , ×, ÷)

Expressions without parentheses that have operations at two levels: first work on operations at the second level (×, ÷) from left to right, then work on operations at the first level (+, -) in order from left to right:

(-20) (4) + 20 = -80 + 20 = - 60 -180 ÷ 9 - 40 = -20 - 40 = -60

- 60 – 40(-2) = -60 + 80 = 20 - 80 + 150 ÷ 3 = - 80 + 50 = - 30

Grade 8 Chapter 1 Lesson 4

Order of Operations in Expression with Parentheses

When expressions have parentheses: at first work on operations in the parentheses, then work on operations before or behind the parentheses:

$3 \times (-90 + 70) = 3(-20) = -60$ $\quad$ $(-50 - 40) \times 2 = -90 \times 2 = -180$

$200 \div (-40 + 20) = 200 \div (-20) = -10$ $\quad$ $(-180 + 40) \div 7 = -140 \div 7 = -20$

Grouping Symbols

Expressions have three kinds of grouping symbols: first work on operations in the parentheses, and then work on operations in the brackets, and last, work on operations in the braces.

Parentheses ( ) $\qquad$ Brackets [ ] $\qquad$ Braces { }

Order of five operations ($+, -, \times, \div, a^n$)

When an expression has five operations: first work on operations at the third level of operations (a^n), then work on operations at second level of operations ($\times, \div$), last, work on operations at first level of operations ($+, -$), in the order, they occur from left to right:

in short

Grouping Symbols	**(1) [2] {3}**
Exponents (Power)	**a^n**
Multiplication or Division from left to right	**× or ÷ L → R**
Addition or Subtract from left to right	**+ or – L → R**

Example 1

$5(-3^2 - 2^2 \cdot 2^3) = 5(-9 - 4 \cdot 8) = 5(-9 - 32) = 5(-41) = -205$

Example 2

$20 - \{-200 \div [-60 - (-30 - 20)]\} = 20 - \{-200 \div [-60 - (-50)]\} =$

$20 - \{-200 \div [-60 + 50]\} = 20 - \{-200 \div (-10)\} = 20 - \{20\} = 20 - 20 = 0$

Grade 8 Chapter 1 Lesson 4

Review 1 A

Perform the indicated operations

1) $-2^4 \div 2^2 + 2$

2) $-3^4 \div 3^2 + 3$

3) $-4^3 \div 2^2 + 2^3$

4) $-6^3 \div 2^2 + 3^3$

5) $-8^3 \div 2^2 + 4^3$

6) $-9^3 \div 3^2 + 3^3$

7) $-3^4 + 3^3 \div 3^2$

8) $-4^3 + 4^2 \div 2^2$

9) $-6^3 + 6^2 \div 3^2$

10) $-8^2 + 6^2 \div 2^2$

11) $-2^2 + 8^2 \div 4^2$

12) $-4^2 + 4^2 \div 2^2$

13) $-5^2 + 6^2 \div 2^2$

14) $-6^2 + 8^2 \div 4^2$

15) $-7^2 + 9^2 \div 3^2$

16) $-8^2 + 6^2 \div 3^2$

17) $-2^3 \div (-4 + 2)^2$

18) $-3^3 \div (-6 + 3)^2$

19) $-4^3 \div (-8 + 4)^2$

20) $-5^3 \div (-9 + 4)^2$

21) $-6^3 \div (-5 + 2)^2$

22) $-8^3 \div (-7 + 3)^2$

23) $-2^5 - 2^4 \div 2^2$

24) $-2^5 - 2^4 \div 2^3$

25) $-2^6 - 2^5 \div 2^4$

26) $-2^6 - 2^5 \div 2^3$

27) $-2^8 \div 2^7 + 2^6$

28) $-2^9 \div 2^8 + 2^7$

29) $-2^9 \div 2^7 + 2^8$

30) $-4^3 \div 2^3 + 1^3$

Grade 8 Chapter 1 Lesson 4

Review 1 B

Perform the indicated operations

1) $-3^4 + 3^3 \div 3^2$

2) $-3^5 + 3^4 \div 3^3$

3) $-3^6 + 3^5 \div 3^4$

4) $-3^3 + 4^3 \div 2^3$

5) $-4^2 - 5^2 - 6^2$

6) $-5^4 + 4^4 + 3^4$

7) $-6^3 + 5^3 + 3^3$

8) $-7^2 - 8^2 + 9^2$

9) $-9^3 + 8^3 - 7^3$

10) $-8^3 + 6^3 - 4^3$

11) $[(-18)^2 + (-16)^2] \div (-10)$

12) $[(-14)^2 - (-12)^2] \div (-52)$

13) $[(-20)^2 - (-19)^2] \div (-13)$

14) $[(-17)^2 - (-15)^2] \div (-8)$

15) $[(-31)^2 - (-29)^2] \div (-3)$

16) $[(-28)^2 - (-25)^2] \div (-3)$

17) $[(-33)^2 - (-29)^2] \div (-8)$

18) $-188 \div [(-24)^2 - (-23)^2]$

19) $[(-5)^2 - (-4)^2]\,[(-3)^2 - (-2)^2]$

20) $[(-9)^2 - (-6)^2] \div [(-8)^2 - (-7)^2]$

21) $-8^2 \div 4^2 + 6^2 \div 3^2$

22) $-9^2 \div 3^2 + 7^2 \div 7$

23) $-10^2 \div 5^2 + 2^3 \div 2^2$

24) $-8^2 \div 2^2 + 9^2 \div 3^2$

25) $-12^2 \div 4^2 + 8^2 \div 4^2$

26) $(-13^2 + 12^2) \div (3^2 - 2^2)$

27) $(-17^2 + 15^2) \div (5^2 - 3^2)$

28) $(-15^2 + 12^2) \div (5^2 - 4^2)$

29) $(-13^2 + 5^2) \div (5^2 - 3^2)$

30) $(-29^2 + 20^2) \div (12^2 - 9^2)$

Grade 8 Chapter 1 Lesson 4
Operations with Rational Numbers

If you can, always in answer try to reduce fractions to lowest term, convert improper fraction to mixed number, and convert fraction to terminating decimal.
If you can, do operations in fractions or decimals but final answers put in decimals. Easy work: + and – in decimals, × and ÷ in fractions.

$-\frac{3}{4}+0.25=-0.75+0.25=-0.5$ $0.5\div(-\frac{1}{8})=\frac{1}{2}\div(-\frac{1}{8})=\frac{1}{2}\times(-\frac{8}{1})=-4$

Evaluate

1) $(-1\frac{3}{8}-1.75)\div 6\frac{1}{4}=(-1\frac{3}{8}-1\frac{3}{4})\div\frac{25}{4}=-2\frac{9}{8}\times\frac{4}{25}=-3\frac{1}{8}\times\frac{4}{25}=-0.5$

2) $(-1.75+1\frac{5}{8})\div\frac{5}{8}=(-1\frac{3}{4}+1\frac{5}{8})\div\frac{5}{8}=(-1\frac{6}{8}+1\frac{5}{8})\div\frac{5}{8}=-\frac{1}{8}\times\frac{8}{5}=-0.2$

3) $(-1\frac{1}{2}-1.8)\div 2.2=(-1\frac{5}{10}-1\frac{8}{10})\div 2\frac{1}{5}=-3\frac{3}{10}\times\frac{5}{11}=-\frac{33}{10}\times\frac{5}{11}=-1.5$

4) $(-2\frac{3}{5}-1.6)\div 2.4=(-2\frac{6}{10}-1\frac{6}{10})\div 2\frac{2}{5}=-3\frac{12}{10}\div\frac{12}{5}=-4\frac{1}{5}\times\frac{5}{12}=-1.75$

5) $1.65\div(-0.6-\frac{1}{2})=1\frac{13}{20}\div(-\frac{3}{5}-\frac{1}{2})=\frac{33}{20}\div(-\frac{6}{10}-\frac{5}{10})=\frac{33}{20}\div(-\frac{11}{10})=-1.5$

6) $2.2\div(-3\frac{2}{5}+1.2)=2\frac{1}{5}\div(-3\frac{4}{10}+1\frac{2}{10})=2\frac{1}{5}\div(-2\frac{2}{10})=2\frac{1}{5}\div(-2\frac{1}{5})=-1$

7) $1\frac{7}{8}\div(-1\frac{1}{5}-1.3)=1\frac{7}{8}\div(-1.2-1.3)=\frac{15}{8}\div(-2.5)=-\frac{15}{8}\times\frac{2}{5}=-0.75$

8) $2.1\div(-4.5+3\frac{1}{3})=2.1\div(-4\frac{3}{6}+3\frac{2}{6})=\frac{21}{10}\div(-1\frac{1}{6})=\frac{21}{10}(-\frac{6}{7})=-1.8$

9) $5\frac{1}{4}\div(-2\frac{1}{4}-1.25)=5\frac{1}{4}\div(-2.25-1.25)=\frac{21}{4}\div(-3.5)=-\frac{21}{4}\times\frac{2}{7}=-1.5$

10) $3.8\div(-4\frac{2}{3}+3.4)=3\frac{4}{5}\div(-4\frac{10}{15}+3\frac{6}{15})=\frac{19}{5}\div(-1\frac{4}{15})=\frac{19}{5}(-\frac{15}{19})=-3$

Grade 8 Chapter 1 Lesson 4
Review 2 A (# 1 - # 15)
Perform the indicated operations

1) $(-1.08+\frac{2}{25})\div\frac{4}{7}$

2) $(-2.5+2\frac{1}{4})\div 1\frac{1}{4}$

3) $(-3\frac{3}{5}+2.4)\div 1\frac{3}{5}$

4) $(-0.5+\frac{3}{4})\div\frac{5}{8}$

5) $(-3\frac{1}{2}+2.4)\div\frac{11}{15}$

6) $(-4\frac{1}{4}+2.125)\div 1\frac{5}{12}$

7) $(-2\frac{1}{2}-3.75)\div 2\frac{1}{2}$

8) $(-3\frac{4}{5}+1.4)\div 2\frac{2}{5}$

9) $(-3\frac{1}{2}+1.7)\div 0.9$

10) $(-3\frac{1}{5}+1.1)\div 4.2$

11) $(-3\frac{1}{2}+2.3)\div 4.8$

12) $(\frac{1}{5}-0.3)\div 0.2$

13) $(-1\frac{1}{2}-2\frac{3}{4})\div 8.5$

14) $(-2\frac{1}{2}+1.4)\div 2.2$

15) $(-3.5+1\frac{1}{6})\div 1\frac{1}{3}$

16) $(-1\frac{1}{2}-1.25)\div 2.2$

Review 2 B (# 16 - # 30)

17) $2.2\div(-2\frac{1}{2}+1.4)$

18) $\frac{8}{25}\div(-0.8+\frac{2}{5})$

19) $5.25\div(-4\frac{7}{10}+2.6)$

20) $0.15\div(-\frac{7}{10}+0.4)$

21) $1.2\div(-2\frac{1}{3}+2\frac{1}{6})$

22) $3.6\div(-2\frac{7}{8}+1\frac{3}{4})$

23) $0.5\div(-\frac{7}{8}+0.75)$

24) $6.75\div(-3\frac{1}{10}-1.4)$

25) $2.2\div(-\frac{1}{2}+0.6)$

26) $2.1\div(-2\frac{2}{5}-1.8)$

27) $\frac{5}{8}\div(-2.25+2\frac{1}{2})$

28) $\frac{7}{20}\div(-0.3+\frac{1}{20})$

29) $3.3\div(-\frac{2}{3}+\frac{1}{6})$

30) $\frac{3}{20}\div(-2.8+2\frac{1}{2})$

Grade 8 Chapter 1 Lesson 5

Laws for Rational Numbers

First Law of Addition (commutative)

Changing the order of addends does not change the sum.
Add in any order:
$A + B = B + A$ $-2 - 3 = -5$ $-3 - 2 = -5$ $-2 - 3 = -3 - 2$

Second Law of Addition (associative)

Changing the grouping of addends does not change the sum:
$A + (B + C) = (A + B) + C$ $-8 + (-2 - 5) = (-8 - 2) - 5 = -10 - 5 = -15$

First Law of Multiplication (commutative)

Changing the order of factors does not change the product.
Multiply in any order:
$A B = B A$ $(-2)(-3) = 6$ $(-3)(-2) = 6$ $(-2)(-3) = (-3)(-2)$

Second Law of Multiplication (associative)

Changing the grouping of factors does not change the product:
$A (B C) = (AB) C$ $-5(-4)(-2) = (-5)(-2)(-4) = 10 \times (-4) = -40$

Distributive Laws of Multiplication

To multiply numbers by the sum or difference, multiply that number by each of the numbers in parentheses, then add or subtract the products:
$A (B + C) = AB + AC$ $A (B - C) = AB - AC$
$(B+C)A = AB + AC$ $(B - C) A = AB - AC$
$-14 \cdot 5 = (-10 - 4) \cdot 5 = -10 \cdot 5 - 4 \cdot 5 = -50 - 20 = -70$
$-18 \cdot 5 = (-20 + 2) \cdot 5 = -20 \cdot 5 + 2 \cdot 5 = -100 + 10 = -90$

Common Factor (Put A before parentheses)

$AB + AC = A (B + C)$ $AB - AC = A (B - C)$
If two products have a common factor, put the common factor before parentheses:
$AB + AC$ A – common factor
In parentheses, the remaining numbers is as result of dividing the product by the common factor: $AB \div A = B$ $AC \div A = C$
$-16 \cdot 5 - 4 \cdot 5 = 5 (-16 - 4) = 5 (-20) = -100$
$-22 \cdot 5 + 2 \cdot 5 = 5 (-22 + 2) = 5 (-20) = -100$
$-24 \times 11 = -24 (10+1) = -24\times10-24\times1 = -240 - 24 = -2$ ($2 + 4 = 6$ – middle digit)

Grade 8 Chapter 1 Lesson 5

Common Factor

If first term negative, put (- A) before parentheses and change the sign before each term: -AB – AC = - A (B + C)   - AB + AC = - A (B - C)

- 16 · 5 - 4 · 5 = - 5 (16 + 4) = - 5 (20) = - 100

- 22 · 5 + 2 · 5 = - 5 (22 - 2) = - 5 (20) = - 100

Examples: Evaluate

1) $-1\frac{3}{5}(2\frac{1}{8})-1\frac{3}{5}(2\frac{7}{8})=-1\frac{3}{5}(2\frac{1}{8}+2\frac{7}{8})=-\frac{8}{5}(5)=-8$

2) $-\frac{3}{4}(4\frac{2}{3})+\frac{3}{4}(3\frac{1}{3})=-\frac{3}{4}(4\frac{2}{3}-3\frac{1}{3})=-\frac{3}{4}(1\frac{1}{3})=-\frac{3}{4}\times\frac{4}{3}=-1$

3) $\frac{5}{6}(-4\frac{3}{5})+\frac{5}{6}\times1\frac{3}{5}=-\frac{5}{6}(4\frac{3}{5}-1\frac{3}{5})=-\frac{5}{6}(3)=-\frac{5}{2}=-2\frac{1}{2}=-2.5$

4) $-4\frac{1}{2}\times\frac{1}{6}-\frac{1}{6}\times1\frac{1}{2}=-\frac{1}{6}(4\frac{1}{2}+1\frac{1}{2})=-\frac{1}{6}(6)=-1$

5) $-3\frac{3}{4}\times1\frac{1}{6}+2\frac{1}{4}\times1\frac{1}{6}=-1\frac{1}{6}(3\frac{3}{4}-2\frac{1}{4})=-1\frac{1}{6}(1\frac{2}{4})=-\frac{7}{6}\times\frac{3}{2}=-\frac{7}{4}=-1.75$

6) $(-\frac{1}{3}+\frac{1}{5})\div\frac{1}{15}=(-\frac{1}{3}+\frac{1}{5})\times15=-\frac{1}{3}\times\frac{15}{1}+\frac{1}{5}\times\frac{15}{1}=-5+3=-2$

7) $(-1\frac{3}{4}+1\frac{1}{2})\div\frac{1}{8}=(-\frac{7}{4}+\frac{3}{2})\times8=-\frac{7}{4}\times\frac{8}{1}+\frac{3}{2}\times\frac{8}{1}=-14+12=-2$

8) $(-1\frac{1}{2}+1\frac{1}{6})\div\frac{1}{6}=(-\frac{3}{2}+\frac{7}{6})\times6=-\frac{3}{2}\times\frac{6}{1}+\frac{7}{6}\times\frac{6}{1}=-9+7=-2$

9) $(-2\frac{1}{2}-1\frac{1}{3})\div\frac{1}{6}=(-\frac{5}{2}-\frac{4}{3})\times6=-\frac{5}{2}\times\frac{6}{1}-\frac{4}{3}\times\frac{6}{1}=-15-8=-23$

10) $(-1\frac{1}{3}-1\frac{3}{4})\div\frac{1}{12}=(-\frac{4}{3}-\frac{7}{4})\times12=-\frac{4}{3}\times\frac{12}{1}-\frac{7}{4}\times\frac{12}{1}=-16-21=-37$

Grade 8 Chapter 1 Lesson 5

Review 1 A

Evaluate by using laws

1) $-\frac{7}{18}-\frac{1}{3}-\frac{5}{18}$

2) $-\frac{3}{14}-\frac{1}{2}-\frac{4}{14}$

3) $-\frac{3}{16}-\frac{3}{4}-\frac{1}{16}$

4) $-\frac{1}{15}-\frac{4}{5}-\frac{2}{15}$

5) $(-\frac{1}{4}-\frac{9}{16})-\frac{7}{16}$

6) $-\frac{7}{8}-\frac{3}{4}-\frac{1}{8}$

7) $-\frac{3}{7}-\frac{5}{9}-\frac{4}{7}-\frac{4}{9}$

8) $(-\frac{1}{7}-\frac{11}{18})+(-\frac{6}{7}-\frac{7}{18})$

9) $-\frac{5}{8}+(-\frac{7}{24}-\frac{3}{8})-\frac{5}{24}$

10) $(-\frac{8}{15}-\frac{1}{5})+(-\frac{4}{15}-\frac{4}{5})$

11) $9(-1+\frac{4}{9})$

12) $(-\frac{2}{3}+\frac{3}{5})\times 15$

13) $9(-3+\frac{5}{9})$

14) $(-1-\frac{5}{6})\times 6$

15) $7(-2-\frac{3}{7})$

16) $(-1+\frac{1}{6})\times 6$

17) $6(-1\frac{1}{2}+\frac{2}{3})$

18) $(-\frac{5}{7}-\frac{4}{7})\times 7$

19) $7(-\frac{5}{7}+\frac{4}{7})$

20) $(-3-\frac{2}{3})\times 3$

21) $3(-2-\frac{1}{6})$

22) $(-2\frac{1}{4}+\frac{5}{8})\times 4$

23) $3(-1\frac{5}{6}+1\frac{1}{3})$

24) $(-3\frac{1}{2}+\frac{1}{4})\times 4$

25) $3(-1\frac{2}{3}+\frac{5}{6})$

26) $(-1\frac{2}{7}+\frac{1}{3})\times 21$

27) $6(-2\frac{1}{2}-1\frac{1}{3})$

28) $(-4-\frac{5}{8})\times 2$

29) $3(-1\frac{1}{3}+\frac{5}{12})$

30) $(-1\frac{1}{4}+\frac{5}{8})\times 4$

Grade 8 Chapter 1 Lesson 5

Review 1 B

Evaluate Using Common Factor

1) $-8\times1\frac{3}{4}+8\times\frac{5}{8}$

2) $-3\frac{2}{9}\times\frac{3}{5}-\frac{3}{5}\times6\frac{7}{9}$

3) $-7\times1\frac{1}{7}+7\times\frac{5}{14}$

4) $-4\times2\frac{1}{4}+4\times1\frac{7}{8}$

5) $-2\frac{6}{7}\times\frac{7}{10}-2\frac{1}{7}\times\frac{7}{10}$

6) $-5\frac{5}{6}\times\frac{3}{4}+1\frac{5}{6}\times\frac{3}{4}$

7) $-4\frac{4}{5}\times\frac{5}{8}-\frac{5}{8}\times3\frac{1}{5}$

8) $-3\frac{2}{3}\times1\frac{1}{2}+1\frac{2}{3}\times1\frac{1}{2}$

9) $-\frac{5}{6}\times5\frac{1}{5}+\frac{5}{6}\times2\frac{1}{5}$

10) $-3\frac{1}{3}\times1\frac{1}{5}-1\frac{1}{5}\times1\frac{2}{3}$

11) $-1\frac{3}{5}\times2\frac{1}{2}-1\frac{3}{5}\times7\frac{1}{2}$

12) $-4\frac{2}{3}\times\frac{3}{4}+3\frac{1}{3}\times\frac{3}{4}$

13) $-4\frac{3}{5}\times\frac{5}{6}+\frac{5}{6}\times1\frac{3}{5}$

14) $-4\frac{1}{2}\times\frac{1}{6}-\frac{1}{6}\times1\frac{1}{2}$

15) $-3\frac{3}{4}\times1\frac{1}{6}+2\frac{1}{4}\times1\frac{1}{6}$

16) $-\frac{1}{3}\times15+\frac{1}{5}\times15$

17) $-1\frac{3}{4}\times8+1\frac{1}{2}\times8$

18) $-1\frac{1}{2}\times6+1\frac{1}{6}\times6$

19) $-2\frac{1}{2}\times6+1\frac{1}{2}\times6$

20) $-1\frac{3}{4}\times12+1\frac{1}{4}\times12$

Evaluate

21) $(-1\frac{3}{5}-1\frac{2}{5})\div\frac{1}{5}$

22) $(-1\frac{3}{5}+1\frac{2}{5})\div\frac{2}{5}$

23) $(-4\frac{3}{5}-1\frac{2}{5})\div\frac{1}{5}$

24) $(-3\frac{2}{3}+1\frac{2}{3})\div\frac{2}{3}$

25) $(-1\frac{3}{4}+1\frac{1}{2})\div\frac{1}{8}$

26) $(-2\frac{1}{2}-1\frac{1}{2})\div\frac{1}{6}$

27) $(-1\frac{4}{5}-1\frac{1}{5})\div\frac{1}{5}$

28) $(-4\frac{3}{5}+1\frac{3}{5})\div\frac{1}{5}$

29) $(-1\frac{1}{3}-1\frac{2}{3})\div\frac{1}{3}$

30) $(-4\frac{1}{5}+2\frac{1}{5})\div\frac{1}{5}$

Grade 8 Chapter 1 Lesson 6
Table 7 Fractions and Decimals
Converting Fractions to Terminating Decimals

$10 = 2\times5$ | $100 = 4\times25$ | $1000 = 8\times125$

$\frac{1}{10} = 0.1$ | $\frac{1}{100} = 0.01$ | $\frac{1}{1000} = 0.001$

$\frac{1}{2} = \frac{1\times5}{2\times5} = \frac{5}{10} = 0.5$ | $\frac{1}{4} = \frac{1\times25}{4\times25} = \frac{25}{100} = 0.25$ | $2\frac{1}{8} = 2\frac{1\times125}{8\times125} = 2\frac{125}{1000} = 2.125$

$\frac{1}{5} = \frac{1\times2}{5\times2} = \frac{2}{10} = 0.2$ | $\frac{15}{20} = \frac{3}{4} = \frac{3\times25}{4\times25} = \frac{75}{100} = 0.75$

Memorize

$\frac{1}{2} = 0.5$ $\quad \frac{1}{4} = 0.25$ $\quad \frac{1}{5} = 0.2$ $\quad \frac{1}{8} = 0.125$ $\quad \frac{3}{4} = 0.75$

Converting Terminating Decimals to Mixed Numbers

a) $1.125 = 1\frac{125}{1000} = 1\frac{125\div5}{1000\div5} = 1\frac{25}{200} = 1\frac{25\div5}{200\div5} = 1\frac{5}{40} = 1\frac{1}{8}$

Combine Operations with Fractions and Decimals

b) $\frac{2}{3}\times0.36 = 0.36\times\frac{2}{3} = 0.36\div3\times2 = 0.12\times2 = 0.24$

c) $3.6\div4\frac{1}{2} = 3.6\div4.5 = 36\div45 = \frac{36}{45} = \frac{4}{5} = 0.8$

d) $(0.66+\frac{31}{50})\div(-\frac{2}{10}) = (\frac{66}{100}+\frac{31}{50})\div(-\frac{1}{5}) = (\frac{33}{50}+\frac{31}{50})(-\frac{5}{1}) = \frac{64}{50}(-\frac{5}{1}) = -6.4$

e) $-0.25\div(\frac{39}{50}-\frac{37}{50}) = -\frac{1}{4}\div\frac{2}{50} = -\frac{1}{4}\times\frac{25}{1} = -\frac{25}{4} = -6\frac{1}{4} = -6.25$

f) $(-\frac{43}{50}+\frac{41}{50})\div0.16 = (-\frac{2}{50})\div\frac{16}{100} = -\frac{1}{25}\div\frac{4}{25} = -\frac{1}{25}\times\frac{25}{4} = -\frac{1}{4} = -0.25$

g) $9.6\div(-\frac{49}{50}-\frac{47}{50}) = 9.6\div(-\frac{96}{50}) = 9\frac{6}{10}\times(-\frac{50}{96}) = 9\frac{3}{5}(-\frac{25}{48}) = -\frac{48}{5}\times\frac{25}{48} = -5$

Easy: + and - in decimals, × and ÷ in fractions

$\frac{1}{2} - 0.25 = 0.5 - 0.25 = 0.25$ $\qquad 0.25\div\frac{1}{8} = \frac{1}{4}\div\frac{1}{8} = \frac{1}{4}\times\frac{8}{1} = 2$

Grade 8 Chapter 1 Lesson 6
Table 8 Rational Numbers

Addition and Subtraction of Rational Numbers

$(-1.25) + (-0.8) = -1.25 - 0.8 = -2.05$ $\quad -1.25 + 0.8 = -0.45$

$-1.25 - (-0.8) = -1.25 + 0.8 = -0.45$ $\quad 0.8 - 1.25 = -0.45$

Multiplication of Rational Numbers

$(+6)(+2) = +12$ $\quad (-6)(-2) = +12$ $\quad (+)(+) = +$ $\quad (-)(-) = +$

$(+6)(-2) = -12$ $\quad (-6)(+2) = -12$ $\quad (+)(-) = -$ $\quad (-)(+) = -$

$(-0.25)(10) = -2.5$ $\quad (0.25)(-100) = -25$ $\quad (-0.25)(-1000) = 250$

$(-4)(2.5) = -10$ $\quad 8(-12.5) = -100$ $\quad (1.25)(-0.8) = -1$

Division of Rational Numbers

$(+6) \div (+2) = +3$ $\quad (-6) \div (-2) = +3$ $\quad (+) \div (+) = +$ $\quad (-) \div (-) = +$

$(-6) \div (+2) = -3$ $\quad (+6) \div (-2) = -3$ $\quad (+) \div (-) = -$ $\quad (-) \div (+) = -$

$(-5) \div 10 = -0.5$ $\quad 5 \div (-100) = -0.05$ $\quad (-5) \div (-1000) = 0.005$

$(-2) \div 5 = -0.4$ $\quad 0.2 \div (-5) = -0.04$ $\quad (-2) \div (-0.5) = 20 \div 5 = 4$

$0.2 \div (-0.5) = (-2) \div 5 = -0.4$ $\quad (-1.2) \div 0.8 = (-12) \div 8 = -1.5$

Powers of Rational Numbers

$2^3 = 2\times2\times2 = 8$ $\quad (-3)^2 = (-3)(-3) = 9$

$(-2)^3 = -2(-2)(-2) = -8$ $\quad (-2)^4 = (-2)(-2)(-2)(-2) = 16$

$-2^3 = -2\times2\times2 = -8$ $\quad -3^2 = -3 \times 3 = -9$

$(-0.1)^2 = (-0.1)(-0.1) = 0.01$ $\quad -0.3^2 = -0.09$

Order of Operations

$-25(-6)(-4) = -25\times4\times6 = -600$ $\quad 20(-4) + 20 = -80 + 20 = -60$

$-60 - 40(-2) = -60 + 80 = 20$ $\quad -16 \div (-4)(-20) = 4(-20) = -80$

$(-50 - 40) \times 2 = -90 \times 2 = -180$ $\quad 3(-90 + 70) = 3(-20) = -60$

$-180 \div 9 - 40 = -20 - 40 = -60$ $\quad -80 + 150 \div 3 = -80 + 50 = -30$

$(-180 + 40) \div 7 = -140 \div 7 = -20$ $\quad 200 \div (-40 + 20) = 200 \div (-20) = -10$

Examples

a) $5(-3^2 - 2^2\times 2^3) = 5(-9 - 4\times8) = 5(-9 - 32) = 5(-41) = -205$

b) $\{-200 \div [-60 - (-30 - 20)]\} = \{-200 \div [-60 + 50]\} = -200 \div (-10) = 20$

Mental Math 9
Law for Decimals

1) $8\times7.5+8\times2.5$ 2) $3.7\times4+4\times1.3$
3) $2.5\times4-1.5\times4$ 4) $5.4\times5-3.4\times5$
5) $7\times3.7+6.3\times7$ 6) $4.9\times2+2\times5.1$
7) $7.5\times6-3.5\times6$ 8) $3.9\times8-8\times1.9$
9) $3\times2.7+3\times2.3$ 10) $2.4\times3+1.6\times3$
11) $9.3 \times 2 - 2 \times 5.3$
12) $5.8 \times 3 + 4.2 \times 3$
13) $4.3 \times 4 + 4 \times 1.7$
14) $9.6 \times 2 - 2.6 \times 2$
15) $5 \times 2.8 - 5 \times 1.8$
16) $3.8 \times 4 - 1.8 \times 4$
17) $2.9 \times 3 + 3.1 \times 3$
18) $0.3 \times 6 + 0.7 \times 6$
19) $6.7 \times 4 - 4 \times 4.7$
20) $8 \times 5.7 - 8 \times 4.7$

Grade 8 Chapter 1 Lesson 6

Mental Math 10

Write the Fractions as Decimals

1) $\frac{3}{10}$

2) $\frac{9}{100}$

3) $\frac{7}{1000}$

4) $\frac{1}{2}$

5) $\frac{12}{24}$

6) $\frac{1}{5}$

7) $\frac{2}{5}$

8) $\frac{3}{5}$

9) $\frac{4}{5}$

10) $\frac{12}{15}$

11) $\frac{1}{4}$

12) $\frac{2}{4}$

13) $\frac{3}{4}$

14) $\frac{3}{12}$

15) $\frac{15}{20}$

16) $\frac{7}{25}$

17) $\frac{7}{50}$

18) $\frac{11}{50}$

19) $\frac{3}{20}$

20) $\frac{9}{20}$

Grade 8 Chapter 1 Lesson 6

Mental Math 11
Negative Fractions

1) $12(-\frac{5}{6})$

2) $-\frac{5}{9}\times 18$

3) $24(-\frac{3}{4})$

4) $-\frac{5}{8}\times 32$

5) $\frac{1}{2}\div(-\frac{1}{2})$

6) $\frac{1}{3}\div(-\frac{1}{6})$

7) $-\frac{1}{4}\div\frac{1}{12}$

8) $-2\div\frac{1}{2}$

9) $(-\frac{1}{2})^2$

10) $(-\frac{1}{5})^2$

11) $(-\frac{2}{5})^2$

12) $-(-\frac{3}{5})^2$

13) $-\frac{1}{4}-\frac{3}{4}$

14) $-\frac{1}{2}-\frac{3}{2}$

15) $-\frac{7}{8}+\frac{5}{8}$

16) $-\frac{7}{8}+\frac{3}{8}$

17) $-\frac{1}{2}-\frac{1}{4}$

18) $-\frac{3}{4}-\frac{1}{2}$

19) $-\frac{1}{2}+\frac{1}{4}$

20) $-\frac{3}{4}+\frac{1}{2}$

Grade 8 Chapter 1 Lesson 6

Mental Math 12
Negative Mixed Numbers

1) $-1-\frac{1}{2}$

2) $-\frac{1}{4}-2$

3) $-1+\frac{1}{2}$

4) $-2\frac{1}{2}+1$

5) $-2\frac{3}{8}-1\frac{1}{8}$

6) $-3\frac{3}{4}-2\frac{1}{4}$

7) $-4\frac{3}{4}+2\frac{1}{4}$

8) $-3\frac{5}{8}+2\frac{1}{8}$

9) $-3\frac{1}{2}-2\frac{1}{2}$

10) $-4\frac{1}{2}-2\frac{1}{2}$

11) $-2\frac{3}{4}+1\frac{1}{4}$

12) $-4\frac{5}{6}+1\frac{2}{6}$

13) $-2\frac{3}{8}+1\frac{1}{8}$

14) $-3\frac{3}{8}+2\frac{1}{8}$

15) $-4\frac{7}{8}+3\frac{3}{8}$

16) $-5\frac{5}{8}+4\frac{1}{8}$

17) $-2+1\frac{1}{2}$

18) $-2+1\frac{3}{4}$

19) $-3+1\frac{1}{4}$

20) $-3+1\frac{1}{5}$

Grade 8 Chapter 1 Lesson 7

Square Roots

The first operation with numbers is addition.
Addition combines numbers together.
Addend + Addend = Sum $3 + 2 = 5$

The second operation with numbers is subtraction.
Subtraction means to take something away.
Minuend – Subtrahend = Difference $5 - 2 = 3$

The third operation with numbers is multiplication.
Multiplication is the addition of the same numbers.
Factor × Factor = Product $3 \times 2 = 6$

The fourth operation with numbers is division.
Division means to separate number by some of equal groups.
Dividend ÷ Divisor = Quotient $6 \div 2 = 3$

The fifth operation with numbers is power.
Power is the multiplication of the same numbers.
Base ^{Exponent} = Power $3^2 = 3 \cdot 3 = 9$

The sixth operation with numbers is root.
The reverse process of squaring is finding a square root.
A square root of number is a number whose square is a number.

$4^2 = 16$ $\sqrt[2]{16} = 4$ **or** $\sqrt{16} = 4$

Perfect square numbers are 4, 9, 16, 25, 36, 49, 64, 81, 100.

$0^2 = 0$ $1^2 = 1$ $2^2 = 4$ $3^2 = 9$ $4^2 = 16$ $5^2 = 25$
$6^2 = 36$ $7^2 = 49$ $8^2 = 64$ $9^2 = 81$ $10^2 = 100$

A square root of perfect square numbers is whole numbers.

$\sqrt{0} = 0$ $\sqrt{1} = 1$ $\sqrt{4} = 2$ $\sqrt{9} = 3$ $\sqrt{16} = 4$ $\sqrt{25} = 5$
$\sqrt{36} = 6$ $\sqrt{49} = 7$ $\sqrt{64} = 8$ $\sqrt{81} = 9$ $\sqrt{100} = 10$

Grade 8 Chapter 1 Lesson 7

Rule for Finding Square Roots from Perfect Square Numbers

The square root of three and four-digit perfect square numbers is always a two-digit number. Divide the number under the root by groups of two digits, from right to left. The last group can be one or two digits.
To find the first digit, find the number whose square is smaller than or equal to the last group.
To find the second digit, use the next rule:

Products $1 \times 1 = 1$ **or** $9 \times 9 = 81$ **have last digit 1.**
Products $2 \times 2 = 4$ **or** $8 \times 8 = 64$ **have last digit 4.**
Products $3 \times 3 = 9$ **or** $7 \times 7 = 49$ **have last digit 9.**
Products $4 \times 4 = 16$ **or** $6 \times 6 = 36$ **have last digit 6.**
Product $5 \times 5 = 25$ **has a last digit 5.**

Last digit of given number	1	4	5	6	9
Last digit of answer	1 or 9	2 or 8	5	4 or 6	3 or 7

Logarithm

The seventh operation with numbers is logarithms

Logarithm = Exponent **If** $b^x = b^y$ **than** $x = y$

$5^x = 125$ $5^x = 5^3$ $x = 3$ $10^x = 100$ $10^x = 10^2$ $x = 2$

Find the exponents

a)	$2^x = 2$	b)	$3^x = 3$	c)	$4^x = 4$	d)	$5^x = 5$
e)	$6^x = 6$	f)	$2^x = 4$	g)	$2^x = 8$	h)	$2^x = 16$
i)	$2^x = 32$	j)	$2^x = 64$	k)	$3^x = 9$	l)	$3^x = 27$
m)	$3^x = 81$	n)	$4^x = 16$	o)	$4^x = 64$	p)	$5^x = 25$
q)	$6^x = 36$	r)	$7^x = 49$	s)	$8^x = 64$	t)	$9^x = 81$

Answers

a)	1	b)	1	c)	1	d)	1
e)	1	f)	2	g)	3	h)	4
i)	5	j)	6	k)	2	l)	3
m)	4	n)	2	o)	3	p)	2
q)	2	r)	2	s)	2	t)	2

Grade 8 Chapter 1 Lesson 7
Find the Square Roots
Review 1 A

Example 1 Find the square root of: $\sqrt{8'41}$

First digit is 2 because: $2^2 = 4$ $3^2 = 9$ $4 < 8$ $9 > 8$

Second digit can be 1 or 9 since the number 841 has a last digit 1, but $21^2 = 441$ and $29^2 = 841$ or $\sqrt{8'41} = 29$

1) $\sqrt{121}$	2) $\sqrt{289}$	3) $\sqrt{169}$	4) $\sqrt{361}$
5) $\sqrt{256}$	6) $\sqrt{144}$	7) $\sqrt{324}$	8) $\sqrt{196}$
9) $\sqrt{441}$	10) $\sqrt{729}$	11) $\sqrt{529}$	12) $\sqrt{841}$
13) $\sqrt{676}$	14) $\sqrt{484}$	15) $\sqrt{784}$	16) $\sqrt{576}$
17) $\sqrt{961}$	18) $\sqrt{1369}$	19) $\sqrt{1089}$	20) $\sqrt{1521}$

Review 1 B

Example 2 Find the square root of: $\sqrt{73'96} = 86$

First digit is 8 because $8^2 = 64$ and $64 < 73$

Second digit can be 4 or 6, check $84^2 = 7056$ and $86^2=7396$

21) $\sqrt{1296}$	22) $\sqrt{1024}$	23) $\sqrt{1444}$	24) $\sqrt{1156}$
25) $\sqrt{1681}$	26) $\sqrt{2209}$	27) $\sqrt{1849}$	28) $\sqrt{2401}$
29) $\sqrt{2116}$	30) $\sqrt{1764}$	31) $\sqrt{2304}$	32) $\sqrt{1936}$
33) $\sqrt{2601}$	34) $\sqrt{3249}$	35) $\sqrt{2809}$	36) $\sqrt{3481}$
37) $\sqrt{3136}$	38) $\sqrt{2704}$	39) $\sqrt{3364}$	40) $\sqrt{2916}$

Grade 8 Chapter 1 Lesson 7
Find the Square Roots

Review 1 C

Example 3 Find $\sqrt{14'44} = 38$ First digit is 3 because $3^2 = 9$ and $9 < 14$
Second digit can be 2 or 8, check $32^2 = 1024$ and $38^2 = 1444$

1) $\sqrt{3721}$	2) $\sqrt{4489}$	3) $\sqrt{3969}$	4) $\sqrt{4761}$
5) $\sqrt{4356}$	6) $\sqrt{3844}$	7) $\sqrt{4624}$	8) $\sqrt{4096}$
9) $\sqrt{5041}$	10) $\sqrt{5929}$	11) $\sqrt{5329}$	12) $\sqrt{6241}$
13) $\sqrt{5776}$	14) $\sqrt{5184}$	15) $\sqrt{6084}$	16) $\sqrt{5476}$
17) $\sqrt{6561}$	18) $\sqrt{7569}$	19) $\sqrt{6889}$	20) $\sqrt{7921}$

Review 1 D

Example 4 Find $\sqrt{22'09} = 47$ First digit is 4 because $4^2 = 16$ and $16 < 22$
Second digit can be 3 or 7, check $43^2 = 1849$ and $47^2 = 2209$

21) $\sqrt{7396}$	22) $\sqrt{6724}$	23) $\sqrt{7744}$	24) $\sqrt{7056}$
25) $\sqrt{8281}$	26) $\sqrt{9409}$	27) $\sqrt{8649}$	28) $\sqrt{9801}$
29) $\sqrt{9216}$	30) $\sqrt{8464}$	31) $\sqrt{9604}$	32) $\sqrt{8836}$
33) $\sqrt{625}$	34) $\sqrt{5625}$	35) $\sqrt{2025}$	36) $\sqrt{9025}$
37) $\sqrt{4225}$	38) $\sqrt{1225}$	39) $\sqrt{7225}$	40) $\sqrt{3025}$

Grade 8 Chapter 1 Lesson 7

Mental Math 13

Exponents

1) 10^2
2) 11^2
3) 12^2
4) 13^2
5) 14^2
6) 15^2
7) 16^2
8) 17^2
9) 18^2
10) 19^2
11) 20^2
12) 21^2
13) 22^2
14) 23^2
15) 24^2
16) 25^2
17) 26^2
18) 27^2
19) 28^2
20) 29^2
21) 30^2
22) 31^2
23) 40^2
24) 50^2
25) 60^2
26) 70^2
27) 80^2
28) 90^2

Mental Math 14
Square Roots

1) $\sqrt{100}$
2) $\sqrt{121}$
3) $\sqrt{144}$
4) $\sqrt{169}$
5) $\sqrt{196}$
6) $\sqrt{225}$
7) $\sqrt{256}$
8) $\sqrt{289}$
9) $\sqrt{324}$
10) $\sqrt{361}$
11) $\sqrt{400}$
12) $\sqrt{441}$
13) $\sqrt{484}$
14) $\sqrt{529}$
15) $\sqrt{576}$
16) $\sqrt{625}$
17) $\sqrt{676}$
18) $\sqrt{729}$
19) $\sqrt{784}$
20) $\sqrt{841}$
21) $\sqrt{900}$
22) $\sqrt{961}$
23) $\sqrt{1600}$
24) $\sqrt{2500}$
25) $\sqrt{3600}$
26) $\sqrt{4900}$
27) $\sqrt{6400}$
28) $\sqrt{8100}$

Grade 8 Chapter 1 Lesson 8

Worksheet 4 A Rational Numbers for Final Exam

Perform the indicated operations

1) $\left(-0.25-\frac{1}{8}\right)\div 1.5$

2) $\left(\frac{1}{16}+\frac{1}{32}\right)\div\left(-\frac{3}{16}\right)$

3) $\left(-0.1-\frac{1}{25}\right)\div 1.4$

4) $\left(\frac{1}{64}+\frac{1}{32}\right)\div\left(-\frac{3}{16}\right)$

5) $\left(-0.05-\frac{1}{40}\right)\div\frac{3}{20}$

6) $\left(0.02+\frac{1}{25}\right)\div\left(-\frac{3}{20}\right)$

7) $\left(-\frac{1}{80}-\frac{1}{40}\right)\div\frac{3}{64}$

8) $\left(0.01+\frac{1}{50}\right)\div\left(-\frac{1}{20}\right)$

9) $\left(-0.01-\frac{1}{25}\right)\div 0.25$

10) $\left(0.01+\frac{1}{20}\right)\div\left(-\frac{12}{20}\right)$

Grade 8 Chapter 1 Lesson 8

Worksheet 4 B Rational Numbers for Final Exam

Perform the indicated operations

1) $\frac{3}{8} \div \left(-0.25 + \frac{1}{8} \right)$

2) $-\frac{5}{32} \div \left(\frac{1}{16} - \frac{1}{32} \right)$

3) $\frac{6}{50} \div \left(-0.1 + \frac{1}{25} \right)$

4) $-\frac{5}{32} \div \left(\frac{1}{32} - \frac{1}{64} \right)$

5) $\frac{3}{20} \div \left(-0.05 + \frac{1}{40} \right)$

6) $-\frac{7}{25} \div \left(\frac{1}{25} - 0.02 \right)$

7) $\frac{13}{40} \div \left(-\frac{1}{40} + \frac{1}{80} \right)$

8) $-\frac{9}{50} \div \left(\frac{1}{50} - 0.01 \right)$

9) $\frac{9}{25} \div \left(-\frac{1}{25} + 0.01 \right)$

10) $-\frac{8}{25} \div \left(\frac{1}{20} - 0.01 \right)$

Grade 8 Chapter 1 Lesson 9
Reviews for Chapter 1
Review 1 A

Date **Name**
Each problem is worth 5 points. **Answers must be in the indicated place.**
Read directions carefully. **Do only 1 operation per 1 step.**
Show all steps for full credits. **Perform the indicated operations.**
Evaluate

1) $\frac{1}{5}+(-2)$

1) ______

2) $-(+1\frac{3}{10})-(-2\frac{7}{10})-(-\frac{1}{5})$

2) ______

3) $(-1\frac{2}{5})(0.35)$

3) ______

4) $(-4.5)\div(+1\frac{1}{2})$

4) ____

5) $(-\frac{2}{5})^3+(-\frac{3}{5})^3$

5) _____

Grade 8 Chapter 1 Lesson 9
Review 1 A

Date **Name**

Each problem is worth 5 points. **Answers must be in the indicated place.**

Read directions carefully. **Do only 1 operation per 1 step.**

Show all steps for full credits. **Perform the indicated operations.**

Evaluate

6) $-4\frac{4}{5} \times \frac{5}{8} - \frac{5}{8} \times 3\frac{1}{5}$

6) ________

7) $3\frac{1}{3} \div (-4\frac{5}{6} + 3\frac{1}{2})$

7) ______

8) $(\frac{1}{5} - 0.3) \div 0.2$

8) ______

9) $0.15 \div (-\frac{7}{10} + 0.4)$

9) ____

10) $\frac{3}{40} \div (-2.8 + 2\frac{1}{2})$

10) _____

Grade 8 Chapter 1 Lesson 9
Review 1 B

Date | **Name**

Each problem is worth 5 points.
Read directions carefully.
Show all steps for full credits.
Answers must be in the indicated place.
Do only 1 operation per 1 step.
Perform the indicated operations.

Evaluate

1) $-4-(-3\frac{2}{5})$

1)________

2) $-\frac{1}{5}-(-\frac{3}{10})-(-2\frac{1}{10})$

2)________

3) $-1\frac{4}{5}(0.45)$

3)________

4) $1.2\div(-1\frac{1}{2})$

4) ______

5) $(\frac{3}{5})^3-(-\frac{2}{5})^3$

5) ______

Grade 8 Chapter 1 Lesson 9

Review 1 B

Date

Name

Each problem is worth 5 points.

Answers must be in the indicated place.

Read directions carefully.

Do only 1 operation per 1 step.

Show all steps for full credits.

Perform the indicated operations.

Evaluate

6) $-3\frac{2}{3}\times 1\frac{1}{2}+1\frac{2}{3}\times 1\frac{1}{2}$

6) ________

7) $1\frac{11}{16}\div(-4\frac{1}{4}+3\frac{1}{8})$

7) ________

8) $(-3\frac{1}{5}+1.1)\div 4.2$

8) ________

9) $\frac{8}{25}\div(-0.8+\frac{2}{5})$

9) ______

10) $\frac{9}{20}\div(-0.3+\frac{1}{20})$

10) ____

Grade 8 Chapter 1 Lesson 10
Tests for Chapter 1
Test 1 A

Date **Name**

Each problem is worth 5 points. **Answers must be in the indicated place.**

Read directions carefully. **Do only 1 operation per 1 step.**

Show all steps for full credits. **Perform the indicated operations.**

Evaluate

1) $\frac{2}{5}+(-3)$

1) ______

2) $-(+2\frac{7}{10})-(-1\frac{3}{10})-(-\frac{3}{5})$

2)______

3) $(-1\frac{3}{5})(0.75)$

3)________

4) $(4.5)\div(-1\frac{1}{4})$

4) ______

5) $-(-\frac{1}{2})^2-(-\frac{1}{5})^2$

5)______

Grade 8 Chapter 1 Lesson 10
Test 1 A

Date

Each problem is worth 5 points.

Read directions carefully.

Show all steps for full credits.

Name

Answers must be in the indicated place.

Do only 1 operation per 1 step.

Perform the indicated operations.

Evaluate

6) $-\frac{5}{6} \times 5\frac{1}{5} + \frac{5}{6} \times 2\frac{1}{5}$

6)________

7) $4\frac{1}{2} \div (-4\frac{5}{6} + 3\frac{1}{3})$

7)________

8) $(\frac{1}{5} - 0.4) \div 0.5$

8)________

9) $0.25 \div (-\frac{8}{10} + 0.3)$

9) _______

10) $5\frac{1}{5} \div (-2.6 + 2\frac{1}{2})$

10) _____

Grade 8 Chapter 1 Lesson 10
Test 1 B

Date **Name**

Each problem is worth 5 points. **Answers must be in the indicated place.**

Read directions carefully. **Do only 1 operation per 1 step.**

Show all steps for full credits. **Perform the indicated operations.**

Evaluate

1) $-5-(-4\frac{2}{5})$

1)________

2) $-\frac{2}{5}-(-\frac{7}{10})-(-2\frac{3}{10})$

2)________

3) $-1\frac{1}{5}(0.25)$

3) _______

4) $1.5\div(-1\frac{1}{4})$

4) _______

5) $(-\frac{1}{5})^2-(-\frac{1}{2})^2$

5)_______

Grade 8 Chapter 1 Lesson 10
Test 1 B

Date
Each problem is worth 5 points.
Read directions carefully.
Show all steps for full credits.

Name
Answers must be in the indicated place.
Do only 1 operation per 1 step.
Perform the indicated operations.
Evaluate

6) $-3\frac{1}{3} \times 1\frac{1}{5} - 1\frac{1}{5} \times 1\frac{2}{3}$

6) _______

7) $1\frac{1}{4} \div (-4\frac{5}{8} + 3\frac{1}{16})$

7) _______

8) $(-2\frac{1}{5} + 0.6) \div 3.2$

8) _______

9) $-\frac{4}{25} \div (-0.6 + \frac{4}{5})$

9) _____

10) $-\frac{9}{20} \div (-0.3 + \frac{7}{20})$

10) ______

Grade 8 Answers of Chapter 1

Lesson 1 Pretest

1) - 6 2) - 0.8 3) - 0.5 4) - 0.8 5) - 9

Lesson 2 Addition, Subtraction of Rational Numbers. Review 1 A

1) - 1	2) 0.5	3) - 9	4) - 6	5) - 3
6) 8	7) - 1.75	8) - 0.5	9) - 1.2	10) 1.5
11) 1.5	12) - 0.5	13) - 4.5	14) - 0.75	15) - 0.6
16) 3.8	17) - 4.8	18) - 0.6	19) - 1.8	20) - 1.8

Review 1 B

1) - 0.375	2) 2.4	3) - 3.5	4) - 1.5	5) - 0.6
6) – 0.1	7) 0.625	8) - 3.125	9) - 0.375	10) - 5.5
11) 0.5	12) 0.5	13) 0.25	14) 1.625	15) - 0.875
16) – 0.5	17) 2.6	18) 2.2	19) 7.25	20) 1.6

Lesson 3 Multiply, Divide, Power of Rational Numbers. Review 1 A

1) 2	2) - 0.25	3) 4	4) - 3
5) - 1	6) - 2	7) - 7	8) 0.5
9) - 1.5	10) - 1.25	11) 1.5	12) - 0.24
13) - 6	14) - 1.6	15) 1.7	16) - 2
17) - 0.5	18) - 0.6	19) 4	20) 0.35

Review 1 B

1) - 2	2) - 10	3) - 4	4) - 0.8
5) - 0.4	6) - 0.4	7) - 1.25	8) 1.5
9) 0.6	10) - 1.25	11) - 2	12) - 0.8
13) - 2.4	14) 3	15) - 2.5	16) - 0.2
17) - 0.12	18) - 0.52	19) – 0.5	20) 1

Lesson 4 Order of Operations of Rational Numbers. Review 1 A

1) - 2	2) - 6	3) - 8	4) -27	5) -64
6) -54	7) -78	8) -60	9) -212	10) -55
11) 0	12) - 20	13) -16	14) -32	15) -40
16) -60	17) -2	18) -3	19) -4	20) -5
21) -24	22) -32	23) -36	24) - 34	25) - 66
26) - 68	27) 62	28) 126	29) 252	30) - 7

Review 1 B

1) - 78	2) - 240	3) -726	4) -19	5) -77
6) -288	7) -64	8) -32	9) -560	10) - 360
11) -58	12) -1	13) -3	14) -8	15) -40
16) -53	17) -31	18) -4	19) 45	20) 3
21) 0	22) - 2	23) - 2	24) - 7	25) - 5
26) - 5	27) - 4	28) - 9	29) - 9	30) - 7

Grade 8 Answers of Chapter 1

Lesson 4 Operations with Rational numbers

Review 2 A

1)	- 1.75	2)	- 0.2	3)	- 0.75	4)	0.4	5)	- 1.5
6)	- 1.5	7)	- 2.5	8)	- 1	9)	- 2	10)	- 0.5
11)	- 0.25	12)	- 0.5	13)	- 0.5	14)	- 0.5	15)	- 2

Review 2 B

16)	- 1.25	17)	- 2	18)	- 0.8	19)	- 2.5	20)	- 0.5
21)	- 7.2	22)	- 3.2	23)	- 4	24)	- 1.5	25)	22
26)	- 0.5	27)	2.5	28)	- 1.4	29)	- 6.6	30)	- 0.5

Lesson 5 Laws for Rational Numbers

Review 1 A

1)	-1	2)	-1	3)	-1	4)	-1	5)	-1.25
6)	-1.75	7)	-2	8)	-2	9)	-1.5	10)	-1.8
11)	-5	12)	-1	13)	-22	14)	-11	15)	-17
16)	-5	17)	-5	18)	-9	19)	-1	20)	-11
21)	-6.5	22)	-6.5	23)	-1.5	24)	- 13	25)	-2.5
26)	-20	27)	-23	28)	-9.25	29)	-2.75	30)	-2.5

Review 1 B

1)	-9	2)	-6	3)	-5.5	4)	-1.5	5)	-3.5
6)	- 3	7)	-5	8)	-3	9)	-2.5	10)	-6
11)	- 16	12)	- 1	13)	- 2.5	14)	- 1	15)	- 1.75
16)	- 2	17)	- 2	18)	- 2	19)	- 6	20)	- 6
21)	- 15	22)	- 0.5	23)	- 30	24)	- 3	25)	- 2
26)	- 24	27)	- 15	28)	- 15	29)	- 9	30)	- 10

Lesson 6 Mental Math 9 Law for Decimals

1)	80	2)	20	3)	4	4)	10	5)	70
6)	20	7)	24	8)	16	9)	15	10)	12
11)	8	12)	30	13)	24	14)	14	15)	5
16)	8	17)	18	18)	6	19)	8	20)	8

Lesson 6 Mental Math 10 Fractions and Decimals

1) 0.3 2) 0.09 3) 0.007 4) 0.5 5) 0.5 6) 0.2 7) 0.4 8) 0.6 9) 0.8 10) 0.8
11)0.25 12)0.5 13)0.75 14) 0.25 15) 0.75 16) 0.28 17) 0.14 18) 0.22 19) 0.15 20) 0.45

Grade 8 Answers of Chapter 1

Lesson 6 Mental Math 11 Negative Fractions

1)	- 10	2)	- 10	3)	- 18	4)	- 20	5)	- 1
6)	- 2	7)	- 3	8)	- 4	9)	0.25	10)	0.04
11)	0.16	12)	- 0.36	13)	- 1	14)	- 2	15)	- 0.25
16)	- 0.5	17)	- 0.75	18)	- 1.25	19)	- 0.25	20)	- 0.25

Lesson 6 Mental Math 12 Negative Mixed Numbers

1)	- 1.5	2)	- 2.25	3)	- 0.5	4)	- 1.5	5)	- 3.5
6)	- 6	7)	- 2.5	8)	- 1.5	9)	- 6	10)	- 7
11)	- 1.5	12)	- 3.5	13)	- 1.25	14)	- 1.25	15)	- 1.5
16)	- 1.5	17)	- 0.5	18)	- 0.25	19)	- 1.75	20)	- 1.8

Lesson 7 Square Roots

Review 1 A

1)	11	2)	17	3)	13	4)	19
5)	16	6)	12	7)	18	8)	14
9)	21	10)	27	11)	23	12)	29
13)	26	14)	22	15)	28	16)	24
17)	31	18)	37	19)	33	20)	39

Review 1 B

21)	36	22)	32	23)	38	24)	34
25)	41	26)	47	27)	43	28)	49
29)	46	30)	42	31)	48	32)	44
33)	51	34)	57	35)	53	36)	59
37)	56	38)	52	39)	58	40)	54

Review 1 C

1)	61	2)	67	3)	63	4)	69
5)	66	6)	62	7)	68	8)	64
9)	71	10)	77	11)	73	12)	79
13)	76	14)	72	15)	78	16)	74
17)	81	18)	87	19)	83	20)	89

Review 1 D

21)	86	22)	82	23)	88	24)	84
25)	91	26)	97	27)	93	28)	99
29)	96	30)	92	31)	98	32)	94
33)	25	34)	75	35)	45	36)	95
37)	65	38)	35	39)	85	40)	55

Grade 8 Answers of Chapter 1

Lesson 7 Mental Math 13 Powers

1)	100	2)	121	3)	144	4)	169
5)	196	6)	225	7)	256	8)	289
9)	324	10)	361	11)	400	12)	441
13)	484	14)	529	15)	576	16)	625
17)	676	18)	729	19)	784	20)	841
21)	900	22)	961	23)	1600	24)	2500
25)	3600	26)	4900	27)	6400	28)	8100

Lesson 7 Mental Math 14 Square Roots

1)	10	2)	11	3)	12	4)	13
5)	14	6)	15	7)	16	8)	17
9)	18	10)	19	11)	20	12)	21
13)	22	14)	23	15)	24	16)	25
17)	26	18)	27	19)	28	20)	29
21)	30	22)	31	23)	40	24)	50
25)	60	26)	70	27)	80	28)	90

Lesson 8 Worksheet 4 A

1)	- 0.25	2)	- 0.5	3)	- 0.1	4)	- 0.25	5)	- 0.5
6)	- 0.4	7)	- 0.8	8)	- 0.6	9)	- 0.2	10)	- 0.1

Worksheet 4 B

1)	- 3	2)	- 5	3)	- 2	4)	- 10	5)	- 6
6)	- 14	7)	- 26	8)	- 18	9)	- 12	10)	- 8

Lesson 9

Review 1 A

1)	- 1.8	2)	1.6	3)	- 0.49	4)	-3	5)	- 0.28
6)	- 5	7)	- 2.5	8)	- 0.5	9)	- 0.5	10)	- 0.25

Review 1 B

1)	- 0.6	2)	2.2	3)	- 0.81	4)	- 0.8	5)	0.28
6)	- 3	7)	- 1.5	8)	- 0.5	9)	- 0.8	10)	- 1.8

Lesson 10

Test 1 A

1)	– 2.6	2)	- 0.8	3)	-1.2	4)	- 3.6	5)	- 0.29
6)	- 2.5	7)	- 3	8)	- 0.4	9)	- 0.5	10)	- 52

Test 1 B

1)	– 0.6	2)	2.6	3)	- 0.3	4)	- 1.2	5)	- 0.21
6)	- 6	7)	- 0.8	8)	- 0.5	9)	- 0.8	10)	- 9

Grade 8

Chapter 2

Equations

Grade 8 Chapter 2 Contents
Equations
Lesson 1 – 10
Basic Problems

Work with lesson 1, read rules and examples, 2) Solve the problems from contents, 3) Check answers in the end of chapter 1, 4) Work with the next lessons 2, 3, 4, 5, …, 5) Solve Test 1 A, 6) Grading Scale: A =10 points, B=9points, C=8 points, D=7 points, F=0-6 points, 7) If you solve 10 problems correctly, you get an A, 8) If have an A, you can move on to next chapter, 9) If your grade is lower than an A, work with the mistakes, then solve Test 1B, 10) When you finish all of the chapters, solve Final Exam in last chapter.

Lesson 1 Solve equation Mental Math Equations. Solve the Pretest 1 # 1 – # 5
Lesson 2 Algebraic expressions. Identities.
Solve Review 2 A # 16 – # 20, Solve Review 2 B # 16 - # 20
Lesson 3 Equations with like terms
Solve Review 1 A # 10, 20, 30, 40 Solve Review 1 B # 36 - # 40
Lesson 4 Finding Unknown
Solve Review 1 A # 5, 10, 15, 20, Solve Review 1 B # 5, 10, 15, 20

Lesson 5 Equations with decimals and fractions
Solve Review 1 A # 36 – # 40, Solve Review 1 B # 36 – # 40
Lesson 6 Conditional equations
Solve Review 1 A # 16 - # 20, Solve Review 1 B # 16 - # 20
Lesson 7 Proportions
Solve Review 1 A # 11 - # 20 odd, Solve Review 1 B # 11 - # 20 odd
Lesson 8 Worksheets 5
Solve Worksheet 5 A # 26 – # 30, Solve Worksheet 5 B # 26 – # 30

Lesson 9 Reviews 2
Solve Review 2 A # 1 - # 10 all, Solve Review 2 B # 1 - # 10 all
Lesson 10 Solve Test 2 A # 1 – # 10 all, Solve Test 2 B # 1 – # 10 all
Answers Chapter 2

Grade 8 Chapter 2 Lesson 1

Mental Math Equations

Solve Equations

1) $x + 6 = 10$
2) $7 + x = 10$

3) $x - 9 = 5$
4) $x - 8 = 9$

5) $x - 7 = 8$
6) $12 - y = 3$

7) $13 - y = 8$
8) $14 - y = 5$

9) $15 - y = 7$
10) $16 - y = 9$

11) $8 \cdot x = 72$
12) $x \cdot 9 = 63$

13) $x \div 8 = 7$
14) $x \div 7 = 6$

15) $x \div 6 = 9$
16) $72 \div z = 9$

17) $72 \div z = 8$
18) $63 \div z = 9$

19) $56 \div z = 7$
20) $54 \div z = 9$

Grades 8 Chapter 2 Lesson 1
Pretest
Solve the equations

1) $0.6p - 0.3p - 0.2 = 2.5$

1)________

2) $\frac{1}{10}g + \frac{2}{5}g + \frac{1}{2} = 1\frac{1}{2}$

2)________

3) $-8(\frac{1}{2} - \frac{1}{4}h) - 3(\frac{2}{3} - \frac{1}{3}h) = 0$

3)________

4) $\frac{x}{4.7} = \frac{2.7}{0.9}$

4) ________

5) $2 \div y = 1\frac{4}{7} \div 3\frac{1}{7}$

5) ______

Grade 8 Chapter 2 Lesson 2
Algebraic expressions

**If Arithmetic is the first part of Mathematics, Algebra is the second part.
Algebra includes operations on variables and solving algebraic equations.
Arithmetic expressions can contain numbers, sign of operations, and parentheses.
$2 + 3 = 5$**

**The value of expression is the result after all operations done.
The value of the expression $2 + 3 = 5$ is 5.
Algebraic expressions can contain numbers, variables, signs of operations, and parentheses.**

**Variables mean letters that represent numbers.
Any numbers can represent letter: $x = 3 \quad 2x = 2(3) = 6$
Addition represented by the words: plus, sum, total, increased by, more than.
Subtraction represented by the words: minus, difference, decreased by and less than, subtract from.**

**Multiplication represented by the words: times, of, product of, twice.
Division represented by the words: divided by, quotient, ratio, and rates.
Some power represented by the words: squared, cubed.
Expressions: Sum $a + b$, Difference $a - b$, Product ab, Quotient $a \div b$**

**Results of comparison can write in equality ($5 = 5$) or inequality:
$2 < 5 \qquad 6 > 4 \qquad 3 \le 7 \qquad 8 \ge 1$
Compare two expressions $a + b$ and $a - b$ if $a = -6 \quad b = 2$
$a + b = -6 + 2 = -4 \quad a - b = -6 - 2 = -8 \quad -4 > -8 \quad a + b > a - b$**

Grades 8 Chapter 2 Lesson 2
Special Algebraic Expressions

The sum of the squares of two terms a and b is the expression $a^2 + b^2$
The difference of the square of two terms a and b is the expression $a^2 - b^2$

The square of the sum of two terms a and b is $(a + b)^2$
The square of the difference of two terms a and b is $(a - b)^2$

The perfect square of the sum of two terms a and b is the expression $a^2 + 2ab + b^2$
The perfect square of the difference of two terms a and b is the expression $a^2 - 2ab + b^2$

The sum of cubes of two terms a and b is the expression $a^3 + b^3$
The difference of cubes of two terms a and b is the expression $a^3 - b^3$

If a = - 6 and b = - 2 find value of expressions a + b, a – b, ab, a ÷ b

$a + b = -6 - 2 = -8$	$a - b = -6 + 2 = -4$
$ab = (-6)(-2) = 12$	$a \div b = (-6) \div (-2) = 3$

Evaluating Algebraic Expressions

$x = -5$	$y = -4$	$x^2 - y^2 = (-5)^2 - (-4)^2 = 25 - 16 = 9$
$a = -3$	$b = -2$	$a^3 - b^3 = (-3)^3 - (-2)^3 = -27 - (-8) = -27 + 8 = -19$
$c = -2$	$d = -1$	$c^2 - 2cd + d^2 = (-2)^2 - 2(-2)(-1) + (-1)^2 = 4 - 4 + 1 = 1$

To compare two arithmetic expressions means to know which value of expressions is biggest or smallest, or if they are equal

Grade 8 Chapter 2 Lesson 2

Identities

Two expressions with the same corresponding value that is equal for any of the variables called identically equal: $a + b = b + a$
If one expression is changed $2(x - 3)$ for another identically equal expression it is called identical transformation of expression: $2x - 6$

Equality that is true for any value of variables called identity.
Equality $a + b = b + a$ or $ab = ba$ is identity

Equality can be identity and can be equation
Identity: $ab = ba$ Equation: $x + 2 = 6$

Identically transformation of expressions uses laws and properties of numbers:

$(-a)(-b) = ab$ $a(-b) = -ab$

Commutative and associative laws of addition and multiplication help simplify expressions

$2 + x + 5 = 2 + 5 + x = 7 + x = x + 7$ $(25)(x)(4) = (25)(4)(x) = 100x$

The distributive law of multiplication used if combining like terms to needed to simplify expressions or to remove parentheses:

$a(b + c) = ab + ac$ $a(b - c) = ab - ac$

$4(x + y) = 4x + 4y$ $-1(t - z) = -t + z$

$(b + c)a = ab + ac$ $(b - c)a = ab - ac$

$(a + b)(-5) = -5a - 5b$ $(m - n)(3) = 3m - 3n$

$-5(x + y) - 3(x - y) = -5x - 5y - 3x + 3y = -8x - 2y$

Remove parentheses

If before a parentheses there is a plus sign or no sign, remove the plus sign and remove the parentheses and rewrite each term with the same sign:

$a + (b - c) = a + b - c$ $(a - b) + (c - d) = a - b + c - d$

If before parentheses there is a minus sign, remove the minus sign, remove the parentheses, and rewrite each term with the opposite sign:

$a - (b - c) = a - b + c$ $(a - b) - (c - d) = a - b - c + d$

Grades 8 Chapter 2 Lesson 2

Review 1 A

Evaluate the value of each algebraic expressions if a = - 5 and b = - 4

1) Sum a and b is a + b. Evaluate: $a + b$
2) Difference a and b is a – b. Evaluate: $a - b$

3) Product a and b is ab. Evaluate: ab
4) Twice of product a and b is 2ab. Evaluate: $2ab$

5) Quotient a and b is a ÷ b. Evaluate: $(a \div b)$
6) Quotient b and a is b ÷ a. Evaluate: $(b \div a)$

7) Square of a is a^2. Evaluate a^2
8) Square of b is b^2. Evaluate b^2

9) Cube of a is a^3. Evaluate a^3
10) Cube of b is b^3. Evaluate b^3

11) Minus square of a is $(-a^2)$. Evaluate $(-a^2)$
12) Minus square of b is $(-b^2)$. Evaluate $(-b^2)$

13) Minus cube of a is $(-a^3)$. Evaluate $(-a^3)$
14) Minus cube of b is $(-b^3)$. Evaluate $(-b^3)$

15) The sum of squares of two terms a and b is the expression $a^2 + b^2$. Evaluate $a^2 + b^2$
16) The difference of squares of two terms a and b is the expression $a^2 - b^2$. Evaluate $a^2 - b^2$

17) The sum of cubes of two terms a and b is the expression $a^3 + b^3$. Evaluate $a^3 + b^3$
18) The difference of cubes of two terms a and b is the expression $a^3 - b^3$. Evaluate $a^3 - b^3$

19) The product of sum of two terms a and b, and the difference of the same terms is $(a + b)(a - b)$. Evaluate $(a + b)(a - b)$
20) Minus twice of product of two terms a and b is $(-2ab)$. Evaluate $(-2ab)$

Grade 8 Chapter 2 Lesson 2

Review 1 B

Compare the value of each algebraic expression if a = - 5 b = - 4

21) **Compare the sum of square of two terms and the difference of square of the same two terms:** $a^2 + b^2$ **and** $a^2 - b^2$

22) **Compare the product of the sum and the difference of two terms and the difference of square of the same terms:**
$(a + b)(a - b)$ **and** $a^2 - b^2$

23) **Compare the square of the difference of two terms and the square of the sum of the same two terms:** $(a - b)^2$ **and** $(a + b)^2$

24) **Compare the sum of cubes of two terms and the difference of cubes of the same two terms:** $a^3 + b^3$ **и** $a^3 - b^3$

25) **Compare the cube of the difference of two terms and the cube of the sum of the same two terms:** $(a - b)^3$ **и** $(a + b)^3$

26) **Compare the square of the difference of two terms and the perfect square of the difference of the same two terms:** $(a - b)^2$ **and** $a^2 - 2ab + b^2$

27) **Compare the square of the sum of two terms and the perfect square of the sum of the same two terms:** $(a + b)^2$ **and** $a^2 + 2ab + b^2$

28) **Compare the perfect square of the sum of two terms and the perfect square of the difference of the same two terms:**
$a^2 + 2ab + b^2$ **and** $a^2 - 2ab + b^2$

29) **Compare the partial perfect square of the difference of two terms and the partial perfect square of the sum of the same two terms:**
$a^2 - ab + b^2$ **and** $a^2 + ab + b^2$

30) **Compare the partial perfect square of the sum of the two terms and the partial perfect of the difference of the same two terms:**
$a^2 + ab + b^2$ **and** $a^2 - ab + b^2$

Grade 8 Chapter 2 Lesson 2
Review 2 A
Simplify the expressions

1) $3(2x + 8) - (5x + 2)$
2) $5(7 - 3x) + 7(2 + 2x)$

3) $8(3 - 2y) + 5(3y + 5)$
4) $3(x - 5) - 2(x - 4)$

5) $5(7y - 2) - 6(5y + 2)$
6) $4(8x + 3) - 6(4x - 3)$

7) $8(2x - 6) - 2(4x + 3)$
8) $3(3y - 4) - 4(2y - 1)$

9) $5(y + 3) - 8(10 - y)$
10) $8(9 + 2x) - 5(2 - 3x)$

11) $-5(0.3y + 1) - 11(y - 1)$
12) $-2(3.1x - 1) - 3(1.2x + 1)$

13) $-5(4.2y + 1) - 5(1.4y - 2)$
14) $-1.6(z - 6) + 1.2(z - 8)$

15) $-0.9(5z - 3) + 0.3(7z + 1)$
16) $0.15(x + 4) - 9.9 - 0.3(x - 1)$

17) $1.6(y - 4) - 0.6 - 3(0.4y - 7)$
18) $3z + 2(2z - 3) - 8 - 7(z - 2)$

19) $10(0.1x - 4) - 100(0.01x - 5)$
20) $100(0.01y - 0.03) - 1000(0.001y - 0.05)$

Grade 8 Chapter 2 Lesson 2

Review 2 B

Simplify the expressions

1) $5(\frac{4}{5}x - 2) - 7(\frac{3}{7}x - 3)$

2) $7(\frac{6}{7}y - 3) - 5(\frac{3}{5}y - 2)$

3) $4(\frac{3}{4} - \frac{1}{4}y) - 3(\frac{1}{3}y + \frac{2}{3})$

4) $5(\frac{2}{5}x - \frac{4}{5}) - 4(\frac{1}{4}x - \frac{3}{4})$

5) $3(\frac{2}{3}x - 2) - 4(\frac{1}{2}x + \frac{3}{4})$

6) $10(\frac{1}{2}y - \frac{3}{5}) - 4(\frac{1}{4}y - \frac{1}{2})$

7) $5(\frac{2}{5}z - 1) - 6(\frac{2}{3}z + \frac{1}{2})$

8) $4(\frac{1}{2}x - \frac{1}{4}y) - 6(\frac{1}{6}x + \frac{1}{3}y)$

9) $5(\frac{2}{5}x + \frac{4}{5}) - 6(\frac{5}{6}x - \frac{2}{3})$

10) $14(\frac{3}{7}y - \frac{5}{7}) - 16(\frac{3}{8}y - \frac{5}{8})$

Grades 8 Chapter 2 Lesson 2

Review 2 B

Simplify the expressions

11) $-6(\frac{1}{6}-\frac{1}{3}x)-2(1-\frac{1}{2}x)$

12) $-8(\frac{1}{8}-\frac{1}{4}y)-3(1-\frac{1}{3}y)$

13) $-5(\frac{2}{5}x+\frac{4}{5})+4(\frac{1}{4}x-\frac{3}{4})$

14) $-\frac{1}{2}(6x-2)-\frac{1}{5}(15x+5)$

15) $-\frac{1}{9}(18y-63)-9(\frac{1}{3}y-\frac{2}{9})$

16) $12(\frac{1}{2}x-\frac{2}{3}y)-\frac{1}{2}(4x-2y)$

17) $16(\frac{3}{4}x+\frac{1}{4}y)-\frac{2}{3}(9x+27y)$

18) $\frac{3}{7}(14x-21y)-15(\frac{2}{3}x-\frac{1}{3}y)$

19) $4(\frac{1}{2}x-\frac{1}{4}y)-\frac{2}{3}(6x+3y)$

20) $\frac{4}{9}(18x-9y)-12(\frac{5}{6}x-\frac{1}{6}y)$

Grade 8 Chapter 2 Lesson 3

Equations

Equality means two expressions joined with an equality sign:
$3 + 2 = 5$
Identity is equality that always true. $A + B = B + A$ $AB = BA$

A formula is an equation that expresses known relationships between two or more variables. Formula: $A + B = B + A$ or $AB = BA$

An equality that contains one letter called an equation with one variable:
$x + 2 = 5$

An expression that stands to the left of the equal sign is called the left side of the equation: $x + 2$

An expression that stands to the right of the equal sign is called the right side of the equation: 5

The value of a variable that makes an equation true equality, is called the solution of the equation.

The solution of an equation is the value of the letter that makes the equation true.

$x + 2 = 5$ $x = 3$ solution $3 + 2 = 5$ True

If the product of factors equals zero, one of the factors equals zero:
$AB = 0$ than $A = 0$ or $B = 0$
$2 \cdot 0 = 0$ $0 \cdot 2 = 0$

Equations having the same solution called equivalent equations.
First equation: $x - 6 = 0$ $x = 6$
Second equation: $2(x - 6) = 0$ $x = 6$
Equations $x - 6 = 0$ and $2(x - 6) = 0$ are equivalent

Grades 8 Chapter 2 Lesson 3

Properties of Equation

1. First Property

To solve an equation, add the same number on both sides of the equation, to obtain equivalent equations

$6 = 6$	$6 + 2 = 6 + 2$	$8 = 8$
$a = b$	$a + c = b + c$	
$x - 2 = 10$	$x - 2 + 2 = 10 + 2$	$x = 12$

2. Second Property

To solve an equation, subtract the same number on both sides of the equation, to obtain equivalent equations

$6 = 6$	$6 - 2 = 6 - 2$	$4 = 4$
$a = b$	$a - c = b - c$	
$x + 2 = 10$	$x + 2 - 2 = 10 - 2$	$x = 8$

3. Third Property

To solve an equation, multiply by the same nonzero number on both sides of the equation, to obtain equivalent equations

Do not multiply by zero.

$6 = 6$	$6 \times 2 = 6 \times 2$	$12 = 12$
$a = b$	$ac = bc$	$c \neq 0$
$\frac{1}{2}x = 10$	$2 \times \frac{1}{2}x = 10 \times 2$	$x = 20$

4. Fourth Property

To solve an equation, divide by the same nonzero number on both sides of the equation, to obtain equivalent equations. Do not divide by zero

$6 = 6$	$6 \div 2 = 6 \div 2$	$3 = 3$
$a = b$	$a \div c = b \div c$	$c \neq 0$
$2\,x = 10$	$2\,x \div 2 = 10 \div 2$	$x = 5$

Grade 8 Chapter 2 Lesson 3

Equations with Like Terms

A term is an expression with numbers, letters, product, and power: $2a^3 b^4$
A number standing before letters in a term is called a coefficient: 2
Terms with the same letters with the same exponent called like terms.

Like terms $3a^2$ **and** $2a^2$ **and unlike terms** $2x^2$ **and** $3y^2$ **or** $2x^2$ **and** $3x^3$
Like terms can have different coefficients: $2b^3$ **and** $4b^3$
To combine or simplify like terms, add their coefficients and multiply by their common letters and exponents:

$3a^2 + 3a^2 = 6a^2$	$2b^3 + 4b^3 = 6b^3$
$3x^2y - 2x^2y = x^2y$	$x + x + x = 3x$
$x + 3x = 4x$	$3x - x = 2x$
$x + 3 + x = x + x + 3 = 2x + 3$	$2 + x + 3 = 2 + 3 + x = 5 + x$
$- 5m - 8n - 4m + 3n =- 9m - 5n$	$2x - 3x = - 1\ x = - x$

Fifth Property of Equations
If both sides of an equation have the same terms with the same sign, drop this term from both sides: $2x + 6 = 6 \quad 2x = 0 \quad x = 0$
If $a + c = b + c$ **than** $a = b$ $\quad 2x + 3 = 4 + 3 \quad 2x = 4 \quad x = 2$
Sixth Property of Equations
If any term is moved from one side of an equation to another side, change sign to opposite sign: $2x - 6 = 0 \quad 2x = 6 \quad x = 3$
If $a + b = c$ **than** $a = c - b$ $\quad 2x + 3 = 7 \quad 2x = 7 - 3 \quad 2x = 4 \quad x = 2$

To solve equations with like terms, simplify or combine like terms.
Example a) $7x + 4x = - 66 \quad 11x = - 66 \quad x = - 6$
Example b) $x + 3x = 60 \quad 4x = 60 \quad x = 15$
Example c) $3x + 8 - x = 36 \quad 3x - x + 8 = 36 \quad 2x + 8 = 36$
$2x = 36 - 8 \quad 2x = 28 \quad x = 14$

Grades 8 Chapter 2 Lesson 3
Review 1 A

Solve the equations, using properties

1) $x - 19 = 11$ (+19)

2) $x - 28 = 12$ (+28)

3) $x - 37 = 13$

4) $x - 46 = 14$

5) $x - 55 = 15$

6) $x - 29 = 12$

7) $x - 38 = 13$

8) $x - 47 = 14$

9) $x - 56 = 15$

10) $x - 65 = 16$

11) $x + 11 = 30$ (-11)

12) $x + 12 = 40$ (-12)

13) $x + 13 = 50$

14) $x + 14 = 60$

15) $x + 15 = 70$

16) $x + 12 = 41$

17) $x + 13 = 51$

18) $x + 14 = 61$

19) $x + 15 = 71$

20) $x + 16 = 81$

Grade 8 Chapter 2 Lesson 3
Review 1 A

Solve the equations, using properties

21) $\frac{1}{2}x = 12\,(2)$

22) $\frac{1}{3}x = 15\,(3)$

23) $\frac{1}{4}x = 16$

24) $\frac{1}{5}x = 25$

25) $\frac{1}{6}x = 30$

26) $\frac{1}{7}x = 35$

27) $\frac{1}{8}x = 40$

28) $\frac{1}{9}x = 45$

29) $\frac{1}{10}x = 50$

30) $\frac{1}{11}x = 55$

31) $0.1y = 11\ (\div\ 0.1)$

32) $0.2y = 12\ (\div 0.2)$

33) $0.3y = 15$

34) $0.4y = 16$

35) $0.5y = 25$

36) $0.6y = 30$

37) $0.7y = 42$

38) $0.8y = 48$

39) $0.9y = 36$

40) $1.1y = 44$

Grades 8 Chapter 2 Lesson 3
Review 1 B

Solve the Equations with Like Terms

1) $2x + 4x = 72$
2) $8y + 4y = 240$
3) $z + 30z = 341$
4) $x + 19x = 2100$

5) $13y - 2y = 44$
6) $9z - 5z = 44$
7) $7x - x = 84$
8) $8y - y = 707$

9) $30z - z = 319$
10) $40x - x = 7956$
11) $y + 2y + 3y = 108$
12) $z + 4z + 5z = 150$

13) $5 + 2 - y = 3$
14) $50 + 20 - z = 30$
15) $13x - 2x + 3 = 14$
16) $8y - 3y - 30 = 15$

17) $25 + 4 + 2z = 29$
18) $75 + 9 + 5x = 1614$
19) $124 - 49 - 2x = 889$
20) $548 - 679 - 3y = 1393$

21) $z + 5 + z = 15$
22) $x + 4 + x = 104$
23) $y + 50 + y = 100$
24) $z + 125 + z = 133$

25) $x + 1234 + x = 568$
26) $2y + 6678 + y = 789$
27) $2 + z + 5 = 17$
28) $1 + x + 30 = 31$

29) $12 + y + 50 = 100$
30) $123 + 2z + 456 = -987$
31) $4x + 28 - x = 136$
32) $15y + 123 - y = 5807$

33) $-234 + 23z = 11427$
34) $-65x - 345 + 9x = 33759$
35) $-6y - 456 - y = 258$
36) $-34z - 567 + 8z = 4711$

37) $-78x - 678 + 9x = 20298$
38) $9y + 789 - 28y = 8484$
39) $28z + 987 - 45z = 9589$
40) $125x - 876 - 301x = 105956$

Grade 8 Chapter 2 Lesson 4

Find the unknowns

Addition

The first number in addition is called the addend: 2
The second number in addition is called the addend: 3
The result of addition is called the sum: 5
Addend + Addend = Sum $2 + 3 = 5$

$x + 2 = 8$	$x + 2 - 2 = 8 - 2$	$\underline{x = 8 - 2}$	$x = 6$
$2 + x = 8$	$2 + x - 2 = 8 - 2$	$\underline{x = 8 - 2}$	$x = 6$

To find one addend, subtract the other addend from the sum.

First Addend = Sum – Second Addend

$x + 3 = 5$ $\quad x = 5 - 3$ $\quad x = 2$

Second Addend = Sum – First Addend

$2 + x = 5$ $\quad x = 5 - 2$ $\quad x = 3$

Subtraction

The first number in subtraction is called the minuend: 5
The second number in subtraction is called the subtrahend: 2
The result of subtraction is called the difference: 3
Minuend – Subtrahend = Difference $5 - 2 = 3$

Finding the minuend:

$x - 2 = 8$ $\quad x - 2 + 2 = 8 + 2$ $\quad \underline{x = 8 + 2}$ $\quad x = 10$

To find minuend, add the difference to the subtrahend

Minuend = Difference + Subtrahend

$x - 2 = 3$ $\quad x = 3 + 2$ $\quad x = 5$

Finding the subtrahend

$8 - x = 2$ $\quad 8 - x - 8 = 2 - 8$ $\quad -x = -6$

$(-1)(-x) = (-1)(-6)$ $\quad x = 6$ $\quad \underline{x = 8 - 2}$

To find the subtrahend, subtract the difference from the minuend.

Subtrahend = Minuend – Difference

$5 - x = 3$ $\quad x = 5 - 3$ $\quad x = 2$

Grade 8 Chapter 2 Lesson 4

Multiplication

The first number in multiplication is called the factor: 2
The second number in multiplication is called the factor: 3
The result of multiplication is called the product: 6
Factor × Factor = Product $2 \times 3 = 6$
Finding the first factor:
$x \cdot 2 = 8$ $x \cdot 2 \div 2 = 8 \div 2$ $\underline{x = 8 \div 2}$ $x = 4$
Finding the second factor:
$2 \cdot x = 8$ $2 \cdot x \div 2 = 8 \div 2$ $\underline{x = 8 \div 2}$ $x = 4$
To find one of the factors, divide the product by the other factor
First Factor = Product ÷ Second Factor
$x \cdot 3 = 6$ $x = 6 \div 3$ $x = 2$
Second Factor = Product ÷ First Factor
$2 \cdot x = 6$ $x = 6 \div 2$ $x = 3$

Division

The first number in division is called the dividend: 8
The second number in division is called the divisor: 4
The result of division is called the quotient: 2
Dividend ÷ Divisor = Quotient $8 \div 4 = 2$
Finding the dividend:
$x \div 2 = 8$ $x \div 2 \cdot 2 = 8 \cdot 2$ $x = \underline{8 \cdot 2}$ $x = 16$
To find the dividend, multiply the quotient by the divisor.
Dividend = Quotient × Divisor
$x \div 2 = 3$ $x = 3 \cdot 2$ $x = 6$
Finding the divisor:
$8 \div x = 2$ $8 \div x \cdot x = 2 \cdot x$ $8 = 2x$
$8 \div 2 = 2 \cdot x \div 2$ $4 = x$ $\underline{8 \div 2 = x}$
To find the divisor, divide the dividend by the quotient.
Divisor = Dividend ÷ Quotient
$6 \div x = 3$ $x = 6 \div 3$ $x = 2$

Grade 8 Chapter 2 Lesson 4
Review 1 A

Find the unknown

1)	$x + 191 = 310$	11)	$532 - y = 393$
2)	$282 + x = 410$	12)	$632 - y = 484$
3)	$x + 373 = 510$	13)	$732 - y = 575$
4)	$464 + x = 610$	14)	$865 - y = 696$
5)	$x + 595 = 754$	15)	$976 - y = 797$
6)	$x - 292 = 129$	16)	$643 - y = 494$
7)	$x - 383 = 138$	17)	$743 - y = 585$
8)	$x - 474 = 147$	18)	$843 - y = 676$
9)	$x - 565 = 156$	19)	$965 - y = 787$
10)	$x - 686 = 168$	20)	$987 - y = 898$

Grades 8 Chapter 2 Lesson 4
Review 1 B

Find the unknown

1) $x \cdot 9 = 918$

2) $8 \cdot x = 1624$

3) $x \cdot 7 = 2128$

4) $6 \cdot x = 2430$

5) $x \cdot 5 = 2530$

6) $x \div 9 = 506$

7) $x \div 8 = 607$

8) $x \div 7 = 708$

9) $x \div 6 = 809$

10) $x \div 5 = 908$

11) $9018 \div z = 9$

12) $16024 \div z = 8$

13) $21028 \div z = 7$

14) $24030 \div z = 6$

15) $25030 \div z = 5$

16) $36042 \div z = 6$

17) $49056 \div z = 7$

18) $64072 \div z = 8$

19) $81072 \div z = 9$

20) $72056 \div z = 8$

Grade 8 Chapter 2 Lesson 5

Equations with Decimals

Examples:

a)	$x + 0.2 = 0.6$	$x = 0.6 - 0.2$	$x = 0.4$
b)	$x - 0.2 = 0.6$	$x = 0.6 + 0.2$	$x = 0.8$
c)	$0.6 - x = 0.2$	$x = 0.6 - 0.2$	$x = 0.4$
d)	$0.2\,x = 0.6$	$x = 0.6 \div 0.2$	$x = 3$
e)	$x \div 0.2 = 0.6$	$x = 0.6\,(0.2)$	$x = 0.12$
f)	$0.6 \div x = 0.2$	$x = 0.6 \div 0.2$	$x = 3$

Review 1 A
Solve the equations

1) $x + 0.2 = 0.8$

2) $0.6 + x = 0.8$

3) $x - 0.2 = 0.6$

4) $0.8 - x = 0.2$

5) $x \cdot 2 = 0.8$

6) $0.4\,x = 0.8$

7) $x \div 0.2 = 0.4$

8) $0.8 \div x = 0.4$

9) $1.2\,x = 6$

10) $1.4\,x = 5.6$

11) $2.4\,x = 6$

12) $12.5\,x = 10$

13) $z - 0.3 = 0.9$

14) $10 - z = 1.23$

15) $z \div 0.3 = 0.4$

16) $0.8 \div z = 0.02$

17) $2.5\,z = 5$

18) $z \cdot 0.8 = 100$

19) $0.2y + 0.4y = 7.2$

20) $0.3y - 0.2y = 4.4$

Grades 8 Chapter 2 Lesson 5

Review 1 A

Solve the equations

21) $0.8z + 0.4z = 2.4$

22) $0.1y + 3y = 3.41$

23) $0.9z - 0.5z = 4.4$

24) $3y - 0.1y = 3.19$

25) $1.7z - z = 8.4$

26) $0.8y - 0.1y = 7.07$

27) $2.5 + 0.4 + y = 2.9$

28) $1.2 + z + 5 = 10$

29) $y + 1.25 + 0.08 = 1.33$

30) $0.1 + y + 3 = 3.1$

31) $0.2 + y + 0.5 = 1.7$

32) $0.5 + 0.2 - z = 0.3$

33) $y + 0.5 + y = 1.5$

34) $0.1y + 5 + 0.1y = 10$

35) $z + 0.4 + z = 10.4$

36) $0.3y - 0.2y - 0.3 = 1.2$

37) $0.8z - 0.3z - 3 = 1.5$

38) $0.6x - 0.3x - 0.2 = 2.5$

39) $0.1y + 0.2\ y + 0.3y = 1.08$

40) $0.1z + 0.4z + 0.5z = 1.5$

Grade 8 Chapter 2 Lesson 5

Equations with Fractions

Examples:

a) $x+\frac{1}{5}=\frac{3}{5}$ $\quad x=\frac{3}{5}-\frac{1}{5}$ $\quad x=\frac{2}{5}$ $\quad x=0.4$

b) $x-\frac{1}{5}=\frac{3}{5}$ $\quad x=\frac{3}{5}+\frac{1}{5}$ $\quad x=\frac{4}{5}$ $\quad x=0.8$

c) $\frac{3}{5}-x=\frac{1}{5}$ $\quad x=\frac{3}{5}-\frac{1}{5}$ $\quad x=\frac{2}{5}$ $\quad x=0.4$

d) $\frac{1}{5}x=\frac{3}{5}$ $\quad x=\frac{3}{5}\div\frac{1}{5}$ $\quad x=\frac{3}{5}\cdot\frac{5}{1}$ $\quad x=3$

e) $x\div\frac{1}{5}=\frac{3}{5}$ $\quad x=\frac{3}{5}\cdot\frac{1}{5}$ $\quad x=\frac{3}{25}$ $\quad x=0.12$

f) $\frac{3}{5}\div x=\frac{1}{5}$ $\quad x=\frac{3}{5}\div\frac{1}{5}$ $\quad x=\frac{3}{5}\cdot\frac{5}{1}$ $\quad x=3$

Review 1 B

Solve the equations

1) $x+\frac{1}{5} = \frac{4}{5}$ $\qquad$ 2) $\frac{3}{5}+x = \frac{4}{5}$ $\qquad$ 3) $x-\frac{1}{5} = \frac{3}{5}$

4) $\frac{4}{5}-x = \frac{1}{5}$ $\qquad$ 5) $x \cdot 2 = \frac{4}{5}$ $\qquad$ 6) $\frac{2}{5}x = \frac{4}{5}$

7) $x\div\frac{1}{5} = \frac{2}{5}$ $\qquad$ 8) $\frac{4}{5}\div x = \frac{2}{5}$ $\qquad$ 9) $1\frac{1}{5}x = 6$

10) $1\frac{2}{5}x = 5\frac{3}{5}$ $\qquad$ 11) $2\frac{2}{5}\ x = 6$ $\qquad$ 12) $12\frac{1}{2}\ x = 10$

13) $2\frac{1}{2}z = 5$ $\qquad$ 14) $z \cdot \frac{4}{5} = 100$ $\qquad$ 15) $z\div\frac{3}{10} = \frac{2}{5}$

16) $\frac{4}{5}\div z = 0.02$ $\qquad$ 17) $z-\frac{3}{10} = 0.9$ $\qquad$ 18) $10-z = 1\frac{23}{100}$

19) $\frac{1}{5}y+\frac{2}{5}y = 7\frac{1}{5}$ $\qquad$ 20) $\frac{1}{10}y+3y = 3.41$

Grade 8 Chapter 2 Lesson 5
Review 1 B

Solve the equations

21) $\frac{4}{5}z + \frac{2}{5}z = 2.4$

22) $0.9z - \frac{1}{2}z = 4.4$

23) $3y - \frac{1}{10}y = 3.19$

24) $\frac{3}{10}y - \frac{1}{5}y = 4\ \frac{2}{5}$

25) $1\frac{7}{10}z - z = 8.4$

26) $\frac{4}{5}y - \frac{1}{10}y = 7\frac{7}{10}$

27) $1\frac{1}{5} + z + 5 = 10$

28) $\frac{1}{10} + y + 3 = 3.1$

29) $2.5 + \frac{2}{5} + y = 2.9$

30) $0.1z + \frac{2}{5}z + \frac{1}{2}z = 1.5$

31) $\frac{1}{2} + \frac{1}{5} - z = 0.3$

32) $\frac{1}{2} + \frac{3}{5} - y = \frac{7}{10}$

33) $\frac{1}{5} + y + \frac{1}{2} = 1\frac{7}{10}$

34) $\frac{1}{10}y + \frac{1}{5}y + \frac{3}{10}y = 1\frac{2}{25}$

35) $y + \frac{1}{2} + y = 1\frac{1}{2}$

36) $\frac{2}{5}z + z = 14$

37) $\frac{1}{10}y + 5 + \frac{1}{10}y = 10$

38) $y + 1\frac{1}{4} + 0.08 = 1.33$

39) $\frac{3}{10}y - \frac{1}{5}y - \frac{3}{10} = 1\frac{1}{5}$

40) $\frac{4}{5}z - 0.3z - 3 = 1.5$

Grade 8 Chapter 2 Lesson 6

Conditional equations

An equation in the form $ax = b$, where x is the variable, a and b are any numbers, is called a linear equation with one variables.
All equations have a left side and a right side.

A conditional equation has one solution: $2x = 8$ $x = 4$
A solution of conditional equation: $ax = b$ if $a \neq 0$ then $x = b \div a$

The value of the variable that makes an equation a true equality called the solution of equation:

$x + 2 = 8$	$x = 6$ Solution	$6 + 2 = 8$ True

Equations having the same solutions called equivalent equations.
To solve an equation means to find all solutions to the equation.

Six properties to solve the equation
if $a = b$ then

1)	$a + c = b + c$	2) $a - c = b - c$		
3)	$ac = b c$	4) $a \div c = b \div c$		$c \neq 0$
5)	$a + c = b + c$	$a = b$		
	$2x + 3 = 4 + 3$	$2x = 4$	$x = 2$	
6)	$a - b = c$	$a = c + b$		
	$2x + 3 = 7$	$2x = 7 - 3$	$2x = 4$	$x = 2$

Solve an equation in 5 steps

$2(x - 1) = 3(x + 2) - 4$	1) Open parentheses $a(b + c) = ab + ac$
$2x - 2 = 3x + 6 - 4$	2) Combine like terms
$2x - 2 = 3x + 2$	3) Eliminate numbers or letter (Addition)
$2x - 2 + 2 = 3x + 2 + 2$	
$2x = 3x + 4$	
$2x - 3x = 4$	4) Move 3x from right to left and put minus 3x
$- x = 4$	
$(- 1)(- x) = (-1)(4)$	5) Multiplication property
$\underline{x = - 4}$	

Grades 8 Chapter 2 Lesson 6
Review 1 A

Solve the equations

1) $3(2x + 8) - (5x + 2) = 0$
2) $-3(3y + 4) + 4(2y - 1) = 0$

3) $5(7 - 3x) + 7(2 + 2x) = 0$
4) $8(3 - 2y) + 5(3y + 5) = 9$

5) $3(x - 5) - 2(x - 4) = 8$
6) $5(7y - 2) - 6(5y + 2) = -24$

7) $4(8x + 3) - 6(4x - 3) = 36$
8) $-5(3y + 1) - 11(y - 1) = -20$

9) $5(x - 7) = 3(x - 4) - 27$
10) $4(y - 3) - 16 = 5(y - 5)$

11) $0.8(2x - 6) = 0.2(4x + 3)$
12) $-1.6(z - 6) = -1.2(z - 8)$

13) $0.8(9 + 2x) = 0.5(2 - 3x)$
14) $0.5(y + 3) = 0.8(10 - y)$

15) $-0.7(5z - 3) = -0.3(7z + 1)$
16) $0.15(x - 4) = 9.9 - 0.3(x - 1)$

17) $1.6(y - 4) - 0.6 = 3(0.4y - 7)$
18) $-2(3.1x - 1) + 3(1.2x + 1) = -14.5$

19) $-5(4.2y + 1) + 4(1.4y - 2) = -20.7$
20) $0.3z + 0.2(2z - 3) = 0.8 - 0.7(z - 2)$

Grade 8 Chapter 2 Lesson 6
Review 1 B

Solve the equations

1) $5(\frac{4}{5}x - 2) - 7(\frac{3}{7}x - 3) = 0$

2) $7(\frac{6}{7}y - 3) - 5(\frac{3}{5}y - 3) = 0$

3) $3(\frac{1}{3}x - 2) - 4(\frac{1}{2}x + \frac{1}{4}) = 0$

4) $10(\frac{1}{2}y - \frac{3}{5}) - 4(\frac{1}{4}y - \frac{1}{2}) = 0$

5) $1.8(\frac{8}{9}z + \frac{1}{6}) - 1.5(\frac{2}{3}z - \frac{1}{5}) = 0$

6) $1.6(\frac{3}{4} - \frac{3}{8}y) - \frac{2}{3}(0.9 - 2.7y) = 0$

7) $\frac{2}{9}(1.8y - 2.7) - 0.9(\frac{1}{3}y - \frac{2}{9}) = 0$

8) $\frac{3}{7}(1.4x - 2.1) - 0.6(\frac{1}{3}x - \frac{1}{6}) = 0$

9) $1.5(\frac{2}{5}x - \frac{4}{5}) - 1.2(\frac{1}{6}x - \frac{2}{3}) = 0$

10) $1.6(\frac{3}{4}y - \frac{5}{8}) - 1.8(\frac{2}{9}y - \frac{7}{9}) = 0$

Grades 8 Chapter 2 Lesson 6
Review 1 B

Solve the equations

11) $\frac{1}{2}(1.2x - 4) - \frac{1}{5}(1.5x + 5) = 0$

12) $\frac{3}{7}(2.8x + 5.6) - 1.5(\frac{2}{3}x - \frac{2}{3}) = 0$

13) $0.4(\frac{1}{2}x - \frac{1}{2}) - \frac{2}{3}(0.6x - 0.6) = 0$

14) $\frac{1}{3}(1.8x - 3.6) - 4.2(\frac{6}{7}x - \frac{1}{7}) = 0$

15) $1.2(\frac{1}{2}x + \frac{2}{3}) - \frac{1}{2}(2x - 6) = 0$

16) $\frac{1}{7}(1.4 - 2.1y) - 1.6(\frac{1}{4} - \frac{1}{8}y) = 0$

17) $-6(\frac{1}{6} - \frac{1}{2}x) - 2(1 - \frac{1}{2}x) = 0$

18) $-8(\frac{1}{2} - \frac{1}{4}y) - 3(\frac{2}{3} - \frac{1}{3}y) = 0$

19) $-8(\frac{1}{4} - \frac{3}{8}y) + 6(\frac{1}{3}y + \frac{1}{2}) = 0$

20) $-5(\frac{2}{5}x + \frac{4}{5}) - 4(\frac{3}{4}x - \frac{1}{4}) = 0$

Grade 8 Chapter 2 Lesson 7

Proportions

The quotient of two numbers called the ratio or rate of these numbers. A ratio shows us how many times a first number is greater than second number, or what portion first number from second number.

Find the ratio of numbers 40 to 20. $40 \div 20 = 2$

Find the ratio of numbers 20 to 40. $20 \div 40 = \frac{1}{2}$

Find the Ratio of Numbers – divide (the same units)

$$\frac{10ft}{10yd} = \frac{10ft}{30ft} = \frac{1}{3} = 1 \div 3$$

Find the Rates of Numbers – divide (different units)

$$\frac{250mi}{10hr} = 25\frac{mi}{hr} = 25mph$$

A special type of equation called a proportion.

Equality of two ratios or rates called proportion

$8 \div 2 = 16 \div 4$ $\frac{8}{2} = \frac{16}{4}$ $4 = 4$ **or** $a \div b = c \div d$ $\frac{a}{b} = \frac{c}{d}$

$a \neq 0$ $b \neq 0$ $c \neq 0$ $d \neq 0$

Numbers a and d called exterior numbers: 8 and 4 or 'a' and 'd'

Numbers b and c called interior numbers: 16 and 2 or 'b' and 'c'

Properties of proportions

If the proportion is true, the product of exterior numbers equals the product of the interior numbers. $a \div b = c \div d$ $a\,d = b\,c$

$8 \div 2 = 16 \div 4$ $8 \cdot 4 = 32$ $16 \cdot 2 = 32$ $8 \cdot 4 = 16 \cdot 2$

If the product of the exterior numbers equals the product of the interior numbers of the proportion, then the proportion is true.

$8 \cdot 4 = 32$ $16 \cdot 2 = 32$ $8 \div 2 = 16 \div 4$ $4 = 4$

If the proportion is true interchange exterior or interior numbers, then new proportions is true too.

$4 \div 2 = 16 \div 8$ $8 \div 16 = 2 \div 4$ $4 \div 16 = 2 \div 8$ **or**

$\frac{4}{2} = \frac{16}{8}$ $2 = 2$ $\frac{8}{16} = \frac{2}{4}$ $\frac{1}{2} = \frac{1}{2}$ $\frac{4}{16} = \frac{2}{8}$ $\frac{1}{4} = \frac{1}{4}$

Grades 8 Chapter 2 Lesson 7

To check proportions means that proportions is correct or not, check cross products is equal or not

$\frac{0.8}{8}=\frac{4}{40}$ **0.8(40)=32** **8·4=32** **yes, proportion is correct**

$\frac{0.2}{4}=\frac{2}{80}$ **80(0.2)=16** **4·2=8** **no, proportion is not correct**

Equations 1

To find one of the exterior numbers of a proportion, divide the product of the two interior numbers by another exterior number:

$\frac{x}{b}=\frac{c}{d}$ or $x \div b = c \div d$ $\quad dx = b\,c$ $\quad x = \frac{bc}{d}$

$\frac{1.2}{3}=\frac{2.4}{x}$ or $1.2 \div 3 = 2.4 \div x$ $\quad 1.2x = 2.4 \cdot 3$ $\quad 1.2x = 7.2$ $\quad x = 6$

Equations 2

To find one of the interior numbers of the proportion, divide the product of the two the exterior numbers by another interior number.

$\frac{a}{b}=\frac{x}{d}$ or $a \div b = x \div d$ $\quad bx = ad$ $\quad x = \frac{ad}{b}$

$\frac{1.2}{x}=\frac{2.4}{6}$ or $1.2 \div x = 2.4 \div 6$ $\quad 2.4\,x = 1.2 \cdot 6$ $\quad 2.4\,x = 7.2$ $\quad x = 4$

Application of proportion: to compare the fractions check cross product:

$\frac{1}{2}=\frac{3}{6}$ $\quad 1 \cdot 6 = 2 \cdot 3$ $\quad 6 = 6$

$\frac{5}{6}<\frac{7}{8}$ $\quad 5 \cdot 8 < 6 \cdot 7$ $\quad 40 < 42$

Solve the proportions

$x \div 3 = 4 \div 2$ $\quad \frac{x}{3}=\frac{4}{2}$ $\quad 2x = 12$ $\quad x = 6$

Solve the equation

$3\frac{1}{3} \div 2\frac{1}{2} = 2\frac{2}{3} \div x$ $\quad 3\frac{1}{3}x = 2\frac{1}{2} \cdot 2\frac{2}{3}$ $\quad x = 2\frac{1}{2} \cdot 2\frac{2}{3} \div 3\frac{1}{3}$

$x = \frac{5}{2} \cdot \frac{8}{3} \div \frac{10}{3}$ $\quad x = \frac{5}{2} \cdot \frac{8}{3} \cdot \frac{3}{10}$ $\quad x = \frac{1 \cdot 4}{1 \cdot 2}$ $\quad x = 2$

Grade 8 Chapter 2 Lesson 7
Review 1 A
Solve the proportions

1) $\frac{24}{72} = \frac{x}{9}$

2) $\frac{60}{45} = \frac{24}{x}$

3) $\frac{42}{x} = \frac{51}{34}$

4) $\frac{18}{12} = \frac{75}{x}$

5) $\frac{69}{46} = \frac{x}{54}$

6) $\frac{4.2}{x} = \frac{5.7}{3.8}$

7) $\frac{x}{5.1} = \frac{2.7}{0.9}$

8) $\frac{x}{4.7} = \frac{27.3}{9.1}$

9) $\frac{7.8}{2.6} = \frac{4.5}{x}$

10) $\frac{4.4}{x} = \frac{8.8}{2.2}$

11) $\frac{y}{5\frac{2}{3}} = \frac{1\frac{7}{9}}{1\frac{7}{27}}$

12) $\frac{z}{\frac{4}{5}} = \frac{3\frac{1}{8}}{1\frac{1}{4}}$

13) $y \div \frac{3}{5} = 3\frac{1}{8} \div 3\frac{3}{4}$

14) $1\frac{7}{8} \div 8\frac{1}{3} = z \div \frac{4}{9}$

15) $3\frac{3}{4} \div 1\frac{1}{8} = 2\frac{2}{3} \div y$

16) $\frac{3}{8} \div \frac{3}{11} = z \div \frac{4}{11}$

17) $y \div \frac{1}{3} = \frac{3}{4} \div \frac{1}{2}$

18) $z \div \frac{3}{14} = 3\frac{1}{9} \div \frac{4}{9}$

19) $y \div 4\frac{1}{6} = \frac{4}{7} \div \frac{20}{21}$

20) $z \div 3\frac{1}{5} = 4\frac{1}{2} \div 2\frac{1}{4}$

Grades 8 Chapter 2 Lesson 7
Review 1 B
Solve the proportions

1) $\frac{3}{14} \div 3\frac{1}{7} = y \div 3\frac{2}{3}$

2) $1\frac{1}{2} \div y = 4\frac{3}{4} \div 2\frac{3}{8}$

3) $y \div \frac{4}{25} = \frac{3}{4} \div \frac{4}{5}$

4) $y \div 2\frac{1}{3} = \frac{4}{7} \div 4\frac{1}{6}$

5) $7\frac{1}{3} \div 2\frac{1}{2} = 3\frac{2}{3} \div y$

6) $6\frac{1}{2} \div 3\frac{1}{4} = 3\frac{1}{2} \div k$

7) $m \div 2\frac{1}{2} = 3\frac{1}{4} \div 6\frac{1}{2}$

8) $n \div 2.7 = 8\frac{1}{3} \div 0.5$

9) $2.8 \div p = 1\frac{2}{3} \div 2\frac{6}{7}$

10) $d \div 6.4 = 1\frac{7}{8} \div 7\frac{1}{2}$

11) $c \div 3\frac{1}{3} = 0.9 \div 0.4$

12) $3\frac{1}{3} \div d = 3\frac{3}{5} \div 2.7$

13) $1\frac{7}{9} \div g = 9\frac{1}{3} \div 2\frac{5}{8}$

14) $h \div \frac{7}{12} = \frac{2}{3} \div \frac{7}{9}$

15) $3\frac{3}{5} \div k = 1\frac{4}{5} \div 1\frac{1}{4}$

16) $2\frac{1}{2} \div 1\frac{1}{2} = 6\frac{2}{3} \div m$

17) $3\frac{3}{5} \div \frac{4}{15} = n \div \frac{5}{9}$

18) $2 \div p = 1\frac{4}{7} \div 3\frac{1}{7}$

19) $5\frac{1}{4} \div 3\frac{1}{2} = 4\frac{1}{2} \div u$

20) $3\frac{1}{3} \div 2\frac{1}{2} = 2\frac{2}{3} \div w$

Grade 8 Chapter 2 Lesson 8

Worksheets 5 for Final Exam

Worksheet 5 A

1) $-2^4 \div 2^2 - 2$

2) $-3^4 \div 3^2 - 3$

3) $-4^3 \div 2^2 - 2^3$

4) $-6^3 \div 2^2 + 3^3$

5) $-8^3 \div 2^2 - 4^3$

6) $-9^3 \div 3^2 - 3^3$

7) $-3^4 - 3^3 \div 3^2$

8) $-4^3 - 4^2 \div 2^2$

9) $-6^3 - 6^2 \div 3^2$

10) $-8^2 - 6^2 \div 2^2$

11) $-2^2 - 8^2 \div 4^2$

12) $-4^2 + 4^2 \div 2^2$

13) $-5^2 - 6^2 \div 2^2$

14) $-6^2 - 8^2 \div 4^2$

15) $-7^2 - 9^2 \div 3^2$

16) $-8^2 - 6^2 \div 3^2$

17) $-4^3 \div (-4 + 2)^2$

18) $-9^2 \div (-6 + 3)^2$

19) $-6^3 \div (-8 + 2)^2$

20) $-5^3 \div (-8 + 3)^2$

21) $-4^3 \div (-6 + 4)^2$

22) $-3^3 \div (-6 + 3)^2$

23) $-2^5 + 2^4 \div 2^2$

24) $-2^5 + 2^4 \div 2^3$

25) $-2^6 + 2^5 \div 2^4$

26) $-2^6 + 2^5 \div 2^3$

27) $-2^8 \div 2^7 - 2^6$

28) $-2^9 \div 2^8 - 2^7$

29) $-2^9 \div 2^7 - 2^8$

30) $-4^3 \div 2^3 - 1^3$

Grades 8 Chapter 2 Lesson 8

Worksheet 5 B for Final Exam

1) $-3^4 - 3^3 \div 3^2$

2) $-3^5 - 3^4 \div 3^3$

3) $-3^6 - 3^5 \div 3^4$

4) $-3^3 - 4^3 \div 2^3$

5) $-4^2 + 5^2 - 6^2$

6) $-5^4 + 4^4 - 3^4$

7) $-6^3 + 5^3 - 3^3$

8) $-7^2 + 8^2 - 9^2$

9) $-9^3 - 7^3 + 8^3$

10) $-8^3 - 6^3 + 4^3$

11) $[(-13)^2 + (-11)^2]\div(-10)$

12) $[(-15)^2 - (-13)^2]\div(-7)$

13) $[(-17)^2 - (-15)^2]\div(-8)$

14) $[(-19)^2 - (-17)^2]\div(-8)$

15) $[(-21)^2 - (-19)^2]\div(-8)$

16) $[(-23)^2 - (-21)^2]\div(-2)$

17) $[(-25)^2 - (-23)^2]\div(-6)$

18) $-1040\div[(-27)^2-(-25)^2)$

19) $[(-6)^2 - (-4)^2][(-4)^2 - (-3)^2]$

20) $[(-8)^2 - (-7)^2]\div[(-3)^2 - (-2)^2]$

21) $-8^2 \div 4^2 - 6^2\div 3^2$

22) $-9^2 \div 3^2 - 7^2 \div 7$

23) $-10^2 \div 5^2 - 2^3 \div 2^2$

24) $-8^2 \div 2^2 - 9^2\div 3^2$

25) $-12^2 \div 4^2 - 8^2 \div 4^2$

26) $(-13^2 + 12^2) \div (-3^2 + 2^2)$

27) $(-17^2 + 15^2) \div (-5^2 + 3^2)$

28) $(-15^2 + 12^2) \div (-5^2 + 4^2)$

29) $(-13^2 + 5^2) \div (-5^2 + 3^2)$

30) $(-29^2 + 20^2) \div (-12^2 + 9^2)$

Grade 8 Chapter 2 Lesson 9
Reviews for Chapter 2
Review 2 A

Date **Name**
Read directions carefully. **Answers must be in the indicated place.**
Show all steps for full credits. **Do only 1 operation per 1 step.**
Each problem is worth 5 points. **Solve 1 problem 5 minutes.**

1) **Simplify the expression**
$1.6(y - 4) - 0.6 - 3(0.4y - 7)$

1)_______

2) **Simplify the expression:**
$4(\frac{1}{2}c - \frac{1}{4}d) - \frac{2}{3}(6c + 3d)$

2) ______

Solve the equations:

3) $\frac{1}{10}k + 10 = 50$

3) ______

4) $81072 \div m = 9$

4) _____

5) $28n - 987 - 45n = 9247$

5) ______

Grades 8 Chapter 2 Lesson 9
Review 2 A

Date | **Name**
Read directions carefully. | **Answers must be in the indicated place.**
Show all steps for full credits. | **Do only 1 operation per 1 step.**
Each problem is worth 5 points. | **Solve 1 problem 5 minutes.**

Solve the equations

6) $0.1p + 0.2p + 0.3p = 1.08$

6) _____

7) $\frac{3}{10}g - \frac{1}{5}g - \frac{3}{10} = 1\frac{1}{5}$

7) ______

8) $-5(\frac{2}{5}h + \frac{4}{5}) - 4(\frac{3}{4}h - \frac{1}{4}) = 0$

8) _____

9) Solve the proportion $\frac{7.8}{3.9} = \frac{4.4}{x}$

9) ______

10) Solve the proportion $5\frac{1}{4} \div 3\frac{1}{2} = 4\frac{1}{2} \div y$

10) ______

Grade 8 Chapter 2 Lesson 9
Review 2 B

Date **Name**

Read directions carefully.
Show all steps for full credits.
Each problem is worth 5 points.

Answers must be in the indicated place.
Do only 1 operation per 1 step.
Solve 1 problem 5 minutes.

1) **Simplify the expression**
$3z + 2(2z - 3) - 8 - 7\,(z - 2)$

1) ______

2) **Simplify the expression**
$\frac{4}{9}(18c - 9d) - 12(\frac{5}{6}c - \frac{1}{6}d)$

2) ______

Solve the equations

3) $\frac{1}{11}k - 5 = 55$

3) ______

4) $72056 \div m = 8$

4) _____

5) $125n - 876 - 201n = 37580$

5) ______

Grades 8 Chapter 2 Lesson 9
Review 2 B

Date | **Name**

Read directions carefully. **Answers must be in the indicated place.**
Show all steps for full credits. **Do only 1 operation per 1 step.**
Each problem is worth 5 points. **Solve 1 problem 5 minutes.**

Solve the equations

6) $0.1p + 0.4p + 0.5p = 1.5$

6) ______

7) $\frac{4}{5}g - 0.3g - 3 = 1.5$

7) ______

8) $1.6(\frac{3}{4}h - \frac{5}{8}) - 1.8(\frac{2}{9}h - \frac{7}{9}) = 0$

8) ______

9) Solve the proportion $\frac{4.4}{z} = \frac{8.8}{2.2}$

9) ______

10) Solve the proportion $3\frac{1}{3} \div 2\frac{1}{2} = 2\frac{2}{3} \div y$

10) ______

Grade 8 Chapter 2 Lesson 10
Tests for Chapter 2
Test 2 A

Date | **Name**
Read directions carefully. | **Answers must be in the indicated place.**
Show all steps for full credits. | **Do only 1 operation per 1 step.**
Each problem is worth 5 points. | **Solve 1 problem 5 minutes.**

1) **Simplify the expression**
$10(0.1x - 4) - 100(0.01x - 5)$

1) ______

2) **Simplify the expression:** $16(\frac{3}{4}c + \frac{1}{4}d) - \frac{2}{3}(9c + 27d)$

2)_______

Solve the equations

3) $\frac{1}{8}k - 10 = 40$

3) ______

4) $49056 \div m = 7$

4) ______

5) $-78n - 678 + 9n = 20298$

5)_____

Grades 8 Chapter 2 Lesson 10
Test 2 A

Date | **Name**

Read directions carefully. **Answers must be in the indicated place.**
Show all steps for full credits. **Do only 1 operation per 1 step.**
Each problem is worth 5 points. **Solve 1 problem 5 minutes.**

Solve the equations

6) $0.8p - 0.3p - 3 = 1.5$

6) ______

7) $\frac{1}{10}g + 5 + \frac{1}{10}g = 10$

7) ______

8) $-6(\frac{1}{6} - \frac{1}{2}h) - 2(1 - \frac{1}{2}h) = 0$

8) _____

9) $\frac{x}{4.1} = \frac{2.7}{0.9}$

9) ______

10) $y \div \frac{7}{12} = \frac{2}{3} \div \frac{7}{9}$

10) ______

Grade 8 Chapter 2 Lesson 10
Test 2 B

Date | **Name**
Read directions carefully. | **Answers must be in the indicated place.**
Show all steps for full credits. | **Do only 1 operation per 1 step.**
Each problem is worth 5 points. | **Solve 1 problem 5 minutes.**

1) **Simplify the expression**
100(0.01y – 0.03) - 1000(0.001 y – 0.05)

1)______

2) **Simplify the expression** $\frac{3}{7}(14c-21d)-15(\frac{2}{3}c-\frac{1}{3}d)$

2) ______

Solve equations

3) $\frac{1}{9}k+15=45$

3) ______

4) $64072 \div m = 8$

4)______

5) $9n + 789 - 28n = 8541$

5)_____

Grades 8 Chapter 2 Lesson 10
Test 2 B

Date | **Name**

Read directions carefully.
Show all steps for full credits.
Each problem is worth 5 points.
Answers must be in the indicated place.
Do only 1 operation per 1 step.
Solve 1 problem 5 minutes.

Solve the equations

6) $0.6p - 0.3p - 0.2 = 2.5$

6)______

7) $\frac{1}{10}g + \frac{2}{5}g + \frac{1}{2} = 1\frac{1}{2}$

7)______

8) $-8(\frac{1}{2} - \frac{1}{4}h) - 3(\frac{2}{3} - \frac{1}{3}h) = 0$

8)______

9) $\frac{x}{4.7} = \frac{2.7}{0.9}$

9) ______

10) $2 \div y = 1\frac{4}{7} \div 3\frac{1}{7}$

10) ______

Grade 8 Answers Chapter 2

Lesson 1 Pretest

1) p =9 2) g = 2 3) h = 2 4) x = 14.1 5) y=4

Lesson 2 Algebraic Expressions

Review 1 A

1) - 9	2) - 1	3) 20	4) 40	5) 1.25
6) 0.8	7) 25	8) 16	9) - 125	10) - 64
11) - 25	12) - 16	13) 125	14) 64	15) 41
16) 9	17) - 189	18) - 61	19) 9	20) - 40

Review 1 B

21) 41 > 9	22) 9 = 9	23) 1 < 81	24) - 189 < - 61	25) - 1 > - 729
26) 1 = 1	27) 81 =81	28) 81 > 1	29) 21 < 61	30) 61 > 21

Lesson 2 Identities

Review 2 A

1)	x + 22	2)	- x + 49	3)	- y + 49	4)	x - 7
5)	5y – 22	6)	8x + 30	7)	8x – 54	8)	y - 8
9)	13y – 65	10)	31x + 62	11)	- 12.5y + 6	12)	- 9.8x - 1
13)	- 28y + 5	14)	- 0.4z	15)	- 2.4z + 3	16)	- 0.15x - 9
17)	0.4y+14	18)	0	19)	460	20)	47

Review 2 B

1)	x + 11	2)	3y - 11	3)	- 2y + 1	4)	x - 1
5)	- 9	6)	4y - 4	7)	- 2z – 8	8)	x - 3y
9)	- 3x + 8	10)	0	11)	3x – 3	12)	3y - 4
13)	- x - 7	14)	- 6x	15)	- 5y + 9	16)	4x – 7y
17)	6x – 14y	18)	- 4x – 4y	19)	- 2x – 3y	20)	- 2x – 2y

Lesson 3 Equations

Review 1 A

1)	30	2)	40	3)	50	4)	60
5)	70	6)	41	7)	51	8)	61
9)	71	10)	81	11)	19	12)	28
13)	37	14)	46	15)	55	16)	29
17)	38	18)	47	19)	56	20)	65
21)	24	22)	45	23)	64	24)	125
25)	180	26)	245	27)	320	28)	405
29)	500	30)	605	31)	110	32)	60
33)	50	34)	40	35)	50	36)	50
37)	60	38)	60	39)	40	40)	40

Grade 8 Answers Chapter 2

Lesson 3 Equations with Like Terms

Review 1 B

1)	12	2)	20	3)	11	4)	105
5)	4	6)	11	7)	14	8)	101
9)	11	10)	204	11)	18	12)	15
13)	4	14)	40	15)	1	16)	9
17)	0	18)	306	19)	- 407	20)	- 508
21)	5	22)	50	23)	25	24)	4
25)	- 333	26)	- 1963	27)	10	28)	0
29)	38	30)	- 783	31)	36	32)	406
33)	507	34)	- 609	35)	- 102	36)	- 203
37)	- 304	38)	- 405	39)	- 506	40)	- 607

Lesson 4 Find the unknown

Review 1 A

1)	119	2)	128	3)	137	4)	146
5)	159	6)	421	7)	521	8)	621
9)	721	10)	854	11)	139	12)	148
13)	157	14)	169	15)	179	16)	149
17)	158	18)	167	19)	178	20)	89

Review 1 B

1)	102	2)	203	3)	304	4)	405
5)	506	6)	4554	7)	4856	8)	4956
9)	4854	10)	4540	11)	1002	12)	2003
13)	3004	14)	4005	15)	5006	16)	6007
17)	7008	18)	8009	19)	9008	20)	9007

Lesson 5 Equations with Decimals

Review 1 A

1)	0.6	2)	0.2	3)	0.8	4)	0.6	5)	0.4
6)	2	7)	0.08	8)	2	9)	5	10)	4
11)	2.5	12)	0.8	13)	1. 2	14)	8.77	15)	0.12
16)	40	17)	2	18)	125	19)	12	20)	44
21)	2	22)	1.1	23)	11	24)	1.1	25)	12
26)	10.1	27)	0	28)	3.8	29)	0	30)	0
31)	1	32)	0.4	33)	0.5	34)	25	35)	5
36)	15	37)	9	38)	9	39)	1.8	40)	1.5

Grade 8 Answers Chapter 2

Lesson 5 Equations with Fractions

Review 1 B

1) 0.6	2) 0.2	3) 0.8	4) 0.6	5) 0.4
6) 2	7) 0.08	8) 2	9) 5	10) 4
11) 2.5	12) 0.8	13) 2	14) 125	15) 0.12
16) 40	17) 1.2	18) 8.77	19) 12	20) 1.1
21) 2	22) 11	23) 1.1	24) 44	25) 12
26) 11	27) 3.8	28) 0	29) 0	30) 1.5
31) 0.4	32) 0.4	33) 1	34) 1.8	35) 0.5
36) 10	37) 25	38) 0	39) 15	40) 9

Lesson 6 Conditional equations

Review 1 A

1) - 22	2) - 16	3) 49	4) 40	5) 15
6) - 0.4	7) 0.75	8) 1	9) - 2	10) - 3
11) 6.75	12) 0	13) - 2	14) 5	15) 1.25
16) 24	17) - 35	18) 7.5	19) 0.5	20) 2

Review 1 B

1) - 11	2) 2	3) - 7	4) 1	5) - 1
6) - 0.5	7) 4	8) 2	9) 1	10) - 0.5
11) 10	12) - 17	13) 1	14) -0.2	15) 9.5
16) -2	17) 0.75	18) 2	19) - 0.2	20) - 0.6

Lesson 7 Proportions

Review 1 A

1) 3	2) 18	3) 28	4) 50	5) 81
6) 2.8	7) 15.3	8) 14.1	9) 1.5	10) 1.1
11) 8	12) 2	13) 0.5	14) 0.1	15) 0.8
16) 0.5	17) 0.5	18) 1.5	19) 2.5	20) 6.4

Review 1 B

1) 0.25	2) 0.75	3) 0.15	4) 0.32	5) 1.25
6) 1.75	7) 1.25	8) 45	9) 4.8	10) 1.6
11) 7.5	12) 2.5	13) 0.5	14) 0.5	15) 2.5
16) 4	17) 7.5	18) 4	19) 3	20) 2

Grade 8 Answers Chapter 2

Lesson 8 Worksheets 5 A for Final Exam

1)	- 6	2)	- 12	3)	- 24	4)	- 81	5)	- 192
6)	- 108	7)	- 84	8)	- 68	9)	- 220	10)	- 73
11)	- 8	12)	- 12	13)	- 34	14)	- 40	15)	- 58
16)	- 68	17)	- 16	18)	- 9	19)	- 6	20)	- 5
21)	- 16	22)	- 3	23)	- 28	24)	- 30	25)	- 62
26)	- 60	27)	- 66	28)	- 130	29)	- 260	30)	- 9

Lesson 8 Worksheet 5 B for Final Exam

1)	- 84	2)	- 246	3)	- 732	4)	- 35	5)	- 27
6)	- 450	7)	- 118	8)	- 66	9)	- 560	10)	- 664
11)	- 29	12)	- 8	13)	- 8	14)	- 9	15)	- 10
16)	- 44	17)	- 16	18)	- 10	19)	140	20)	3
21)	- 8	22)	- 16	23)	- 6	24)	- 25	25)	- 13
26)	5	27)	4	28)	9	29)	9	30)	7

Lesson 9 Review 2 for Chapter 2

Review 2 A

1) 0.4y + 14 2) - 2c – 3d 3) k = 400 4) m = 9008 5) n = - 602
6) p = 1.8 7) g = 15 8) h = - 0.6 9) x=2.2 10) x = 3

Review 2 B

1) 0 2) - 2c – 2d 3) k = 660 4) m = 9007 5) n =- 506
6) p = 1.5 7) g = 9 8) y= - 0.5 9) z = 1.1 10) y = 2

Lesson 10 Test 2 for Chapter 2

Tests 2 A

1) 460 2) 6c – 14d 3) x = 400 4) m = 7008 5) n = - 304
6) p = 9 7) g=25 8) h = 0.75 9) x=12.3 10) y=0.5

Test 2 B

1) 47 2) - 4c – 4d 3) k = 270 4) m = 8009 5) n = - 408
6) p =9 7) g = 2 8) h = 2 9) x = 14.1 10) y=4

Grade 8

Chapter 3

Geometry

Grade 8 Chapter 3 Contents
Geometry
Lesson 1 – 12
Basic Problems

Work with lesson 1, read rules and examples, 2) Solve the problems from contents, 3) Check answers in the end of chapter 1, 4) Work with the next lessons 2, 3, 4, 5, …, 5) Solve Test 1 A, 6) Grading Scale: A =10 points, B=9points, C=8 points, D=7 points, F=0-6 points, 7) If you solve 10 problems correctly, you get an A, 8) If have an A, you can move on to next chapter, 9) If your grade is lower than an A, work with the mistakes, then solve Test 1B, 10) When you finish all of the chapters, solve Final Exam in last chapter.

Grade 8 Chapter 3 Lesson 1
Table 12 Geometry

Right triangles

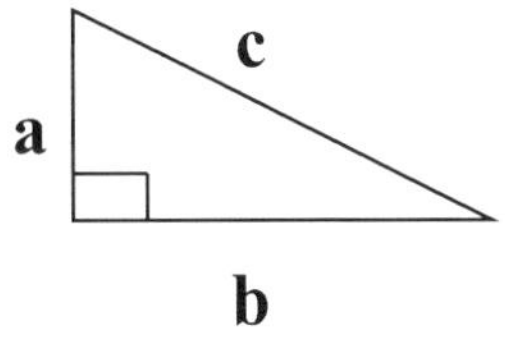

1) Pythagorean Theorem — 1) $a^2 + b^2 = c^2$
2) Area — 2) $A_\Delta = 0.5ab$

Triangles

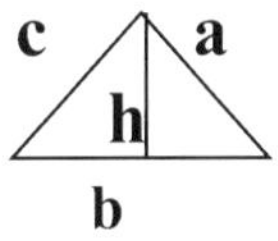

3) Perimeter — 3) $P = a + b + c$
4) Area — 4) $A_\Delta = 0.5bh$

Parallelogram

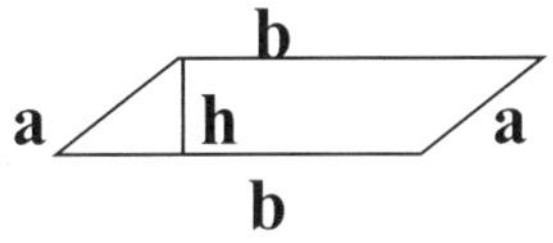

5) Perimeter — 5) $P_{par} = 2a + 2b$
6) Area — 6) $A_{par} = b\ h$

Rhombus

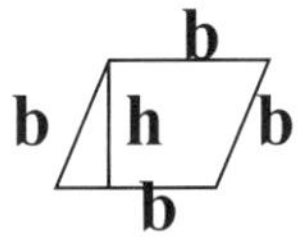

7) Perimeter — 7) $P_{rh} = 4b$
8) Area — 8) $A_{rh} = b \cdot h$

Rectangle

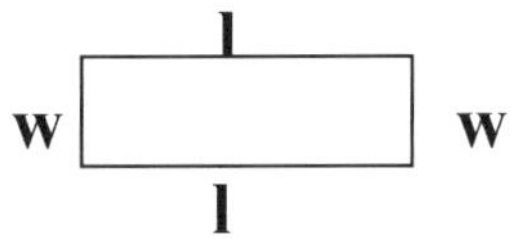

9) Perimeter — 9) $P_{rec} = 2l + 2w$
10) Area — 10) $A_{rec} = l \cdot w$

Square

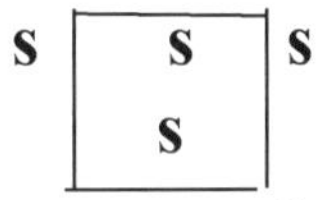

11) Perimeter — 11) $P_{\blacksquare} = 4s$
12) Area — 12) $A_{\blacksquare} = s^2$

Trapezoid

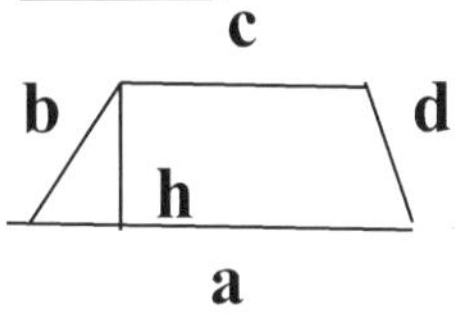

13) Perimeter — 13) $P_{tr} = a + b + c + d$
14) Area — 14) $A_{tr} = 0.5\ h\ (a + c)$

Circle

15) Number π in fraction $\approx \frac{22}{7}$ 16) Number π in decimal ≈ 3.14

17) Diameter $d = 2r$ 18) Circumference 1 (d) $C = \pi d$

19) Circumference 2 (r) $C = 2\pi r$ 20) Area $A_c = \pi r^2$

Solid Geometry

Cube 1) $V = s^3$ 2) $SA = 6s^2$ 3) $d = s\sqrt{3}$ 4) Sphere $V = \frac{4}{3}\pi r^3$ 5) $SA = 4\pi r^2$

Rectangular Solid: 6) $V = lwh$ 7) $SA = 2lw + 2lh + 2wh$ 8) $d^2 = l^2 + w^2 + h^2$

Regular Prism 9) $V = \text{Base} \cdot h$ Regular Pyramid 10) $V = \frac{1}{3}\text{Base} \cdot h$

Right Cylinder 12) $V = \pi r^2 h$ Right Cone 12) $V = \frac{1}{3}\pi r^2 h$

Memorize Formulas of Geometry

Grade 8 Chapter 3 Lesson 1
Pretest
Solve the problems

1) Find the area of a trapezoid, if its height is $3\frac{1}{3}$ cm, and the bases are $9\frac{3}{4}$ cm and $8\frac{1}{4}$ cm. Draw the picture.

1) ______

2) Find the area of a circle if the radius is 14 in and $\pi = \frac{22}{7}$.

Draw the picture.

2) ______

3) Find the area of a ring of a circle if its radii are 6.5 in and 5.5 in and $\pi \approx 3.14$. Draw the picture.

3) ______

4) Find the volume of a rectangular solid if its length is 12.5 in, its width is 5 in and its height is 4 in

4) ______

5) Find the exact value volume of a sphere if the radius r = 6 ft.

5) ______

Grade 8 Chapter 3 Lesson 2

Segments

Geometry is the study of properties of figures with different shapes and sizes. Geometry has two parts: plane and solid geometry.

Plane geometry is the study of properties of flat figures in the plane.

A plane is a flat surface like a desk or a blackboard.

A flat figure is contained in a plane.

Solid geometry is the study of properties of figures having three dimensions.

A point is like a tiny dot and represented by capital letter A, B, C, …

A straight line goes off in both directions, the line is endless.

A straight line divides a plane by semi planes.

A straight line represented by two capital letters or one small letter.

A straight line constructed with a ruler and formed by two points.

Figure 1

Two lines with one common point called intersection lines.

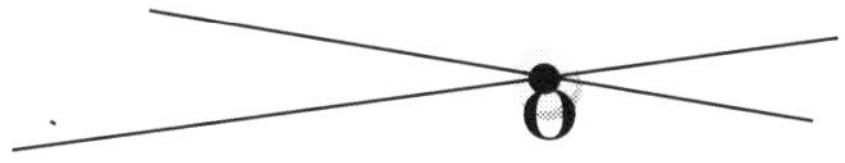

Figure 2

Grade 8 Chapter 3 Lesson 2
Two lines, Rays, Segments

Two lines on the plane called parallel if they have no common point intersection.

Symbol of parallel lines is a || b

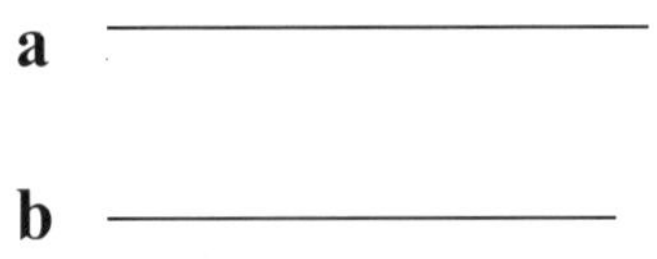

Figure 3
Part of the line limited by one point called a ray.

A ray represented by two capital letters OC.

The first letter is the origin of the ray; the second letter is any point on the ray.

Figure 4
The part of a line limited by two points and all points between the two points called a straight segment.

A segment has an origin and a point that called an end.

A segment is represented by two capital letters and has a positive length AB > 0

Figure 5

Grade 8 Chapter 3 Lesson 2

Measurement

Segment measures in units of length in the American system:
in, ft, yd, and mi

1 inches = 1 in	**1 foot = 1 ft**	**1 yard = 1 yd**	**1 mile = 1 mi**
1 ft = 12 in	**1 yd = 3 ft**	**1 yd = 36in**	**1 mi = 5280 ft**

Segment measures in units of length in the Metric system:
mm, cm, dm, m, and km

1cm = 10mm	**1dm = 10cm**	**1m = 10dm**	**1dm = 100mm**
1m = 100cm	**1m = 1000mm**	**1km = 1000m**	

Segments measured and constructed with a ruler that has origin and units.

Two segments are parallel if they lie on the parallel lines.

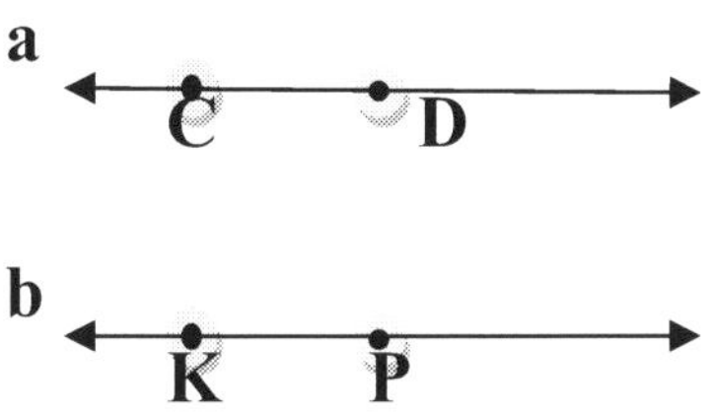

CD = 2 cm **a || b**

KP = 20 mm **CD ||KP**

Figure 6

Grade 8 Chapter 3 Lesson 2
Operations with segments

**Two figures are congruent, if put over of each other, they coincide.
Congruent segments have the same lengths.
Smallest segment has smallest length.
To compare two segments means to find which segments are more or less than another or maybe its equals.**

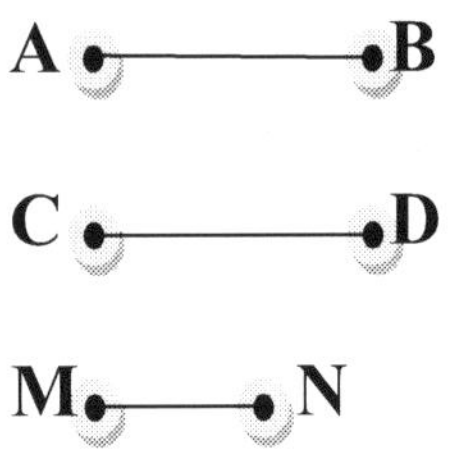

AB = 4 cm	**AB = CD**
CD = 40 mm	**MN< AB**
MN = 3 cm	**CD > MN**

Figure 7

When a point divides a given segment into two segments, then the length of given segment is equal to the sum of the lengths of the two segments. $AC = AB + BC$

Segments can added and subtracted.

$AB + BC = AC$ $AC - AB = BC$ $AB = 8$ **in** $BC = 2$ **in**

$AC = AB + BC = 8 \text{ in} + 2 \text{ in} = 10 \text{ in}$

$BC = AC - AB = 10 \text{ in} - 8 \text{ in} = 2 \text{ in}$

Figure 8

When a point divides a segment into two equal segments or by half, this point called the middle of segment. $AB = BC$.

Segments can multiply or divide by number:

$AB \times 2 = AC$ $AC \div 2 = AB$ $AB = BC = 5$ **in**

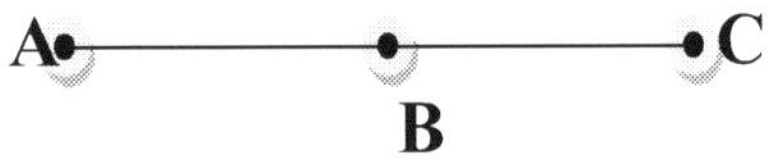

$AC = AB \times 2 = 5 \times 2 = 10 \text{ in}$

$AB = AC \div 2 = 10 \div 2 = 5 \text{ in}$

Figure 9

Grade 8 Chapter 3 Lesson 2
Segments
Review 1 A
Solve the problems
Draw the pictures

1) If AB = 6cm   BC = 4cm   MN = 40mm
Compare segments
a) AB and BC   b) BC and AB   c) BC and MN

2) Add and subtract segments, if AB = 6cm   BC = 4cm
a) AB + BC   b) AB – BC

3) Find AB + BC if AB = 8 cm   BC = 4cm   Draw a picture

4) Find DE + EF if DE = 6 in   EF = 2 in   Draw a picture

5) Find AB – BC if AB = 7in   BC = 5in

6) Find DE – EF if DE = 8 in   EF = 3 in

7) Find AB × 3 if AB = 2cm   Draw a picture

8) Find 4 × DE if DE = 2 in

9) Find AB ÷ 3 if AB = 12cm   Draw a picture

10) Find DE ÷ 2 if DE = 14 in

Grade 8 Chapter 3 Lesson 2
Review 1 B
Solve the problems
Draw the pictures

1) Find the sum of two segments if the lengths of segments equal 3 in and 7 in

2) Find the sum of two segments if the lengths of two segments equal, 1 dm and 5 cm 1 dm = 10 cm

3) How much longer is segment AB than segment CD if AB = 10 in and CD = 4 in?

4) How much shorter is segment CD than segment AB if AB = 10 cm and CD = 4 cm?

5) Find the second segment if the length of the first segments is 5 in and the length of the second segment is 2 in more than first segment.

6) Find the second segment if the length of the first segment is 8in and the length of the second segment is 3 in less than the length of first segment.

7) How much longer is the first segment than the second segment if the lengths of the segments are 1 dm and 4 cm?
1 dm = 10cm

8) How much shorter is the second segment than the first segment if the length of segments are 6 cm and 1 dm. 1 dm = 10 cm

9) Find the sum of two segments if the lengths of the two segments are 5 in and 3 in

10) Find the difference of two segments if the lengths of two segments are 6 in and 3 in

Grade 8 Chapter 3 Lesson 3
Angles

An angle in a figure formed when two rays meet at a common point.
The common point called the vertex.
The rays called the sides of the angle.

An angle represented by three capital letters or one letter on vertex.

Any angle divides a plane by two parts.
An arc denotes the inner part of the angle.

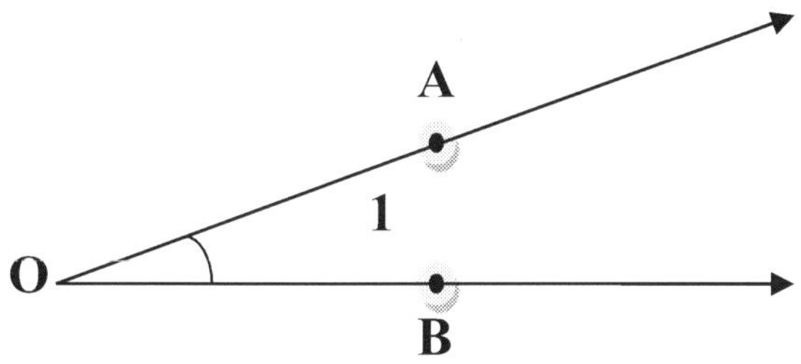

$\angle AOB$ $\angle BOC$
$\angle O$ $\angle 1$

Figure 1

Angles measured in degrees, minutes, and seconds:
1 degree = 60minutes $1^0 = 60'$ **1 minute = 60 seconds** $1' = 60''$

A protractor is an instrument used for measurement and construction of angles.
Each angle has a positive measurement: $\angle ABC > 0$

Congruent angles have the same measurement:

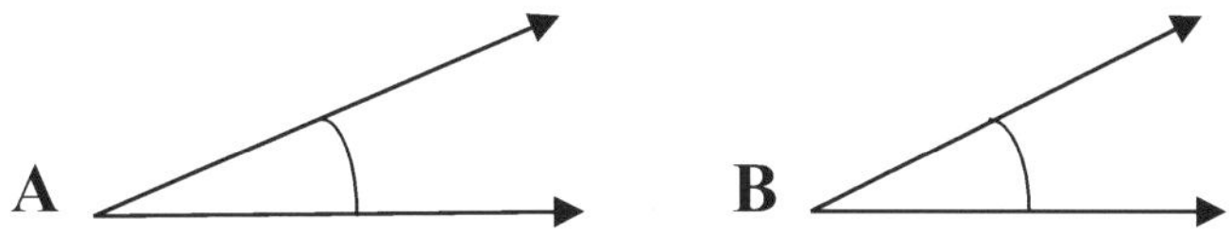

$\angle A = 30^0$
$\angle B = 30^0$
$\angle A = \angle B$

Figure 2

Grade 8 Chapter 3 Lesson 3
Operations with angles

A ray starting from the vertex of an angle and dividing this angle into two equal parts called the bisector of an angle and is denoted: OB = l

Angles can added and subtracted:

$\angle AOB + \angle BOC = \angle AOC$ $\quad$ $\angle AOC - \angle AOB = \angle BOC$

Angles can multiply and divide by number:

$2 \times \angle AOB = \angle AOC$ $\quad$ $\angle AOC \div 2 = \angle AOB$

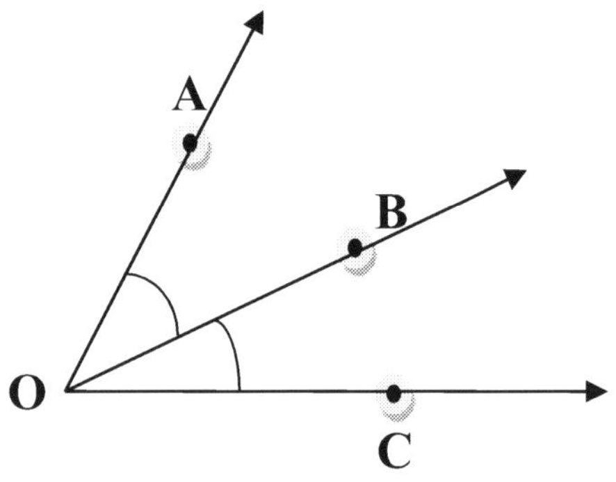

$\angle AOB = 30^0$

$\angle BOC = 30^0$

Figure 3

$\angle AOC > \angle AOB$
$\angle BOC < \angle AOC$

$\angle AOC = \angle AOB + \angle BOC = 30^0 + 30^0 = 60^0$
$\angle BOC = \angle AOC - \angle AOB = 60^0 - 30^0 = 30^0$

$\angle AOC = 2 \times \angle AOB = 2 \times 30^0 = 60^0$
$\angle AOB = \angle AOC \div 2 = 60^0 \div 2 = 30^0$

Grade 8 Chapter 3 Lesson 3
Different kind of angles

An angle larger than 0^0 but smaller than 90^0 is an acute angle:

$0^0 \leq \angle A < 90^0$ **$\angle A = 30^0$**

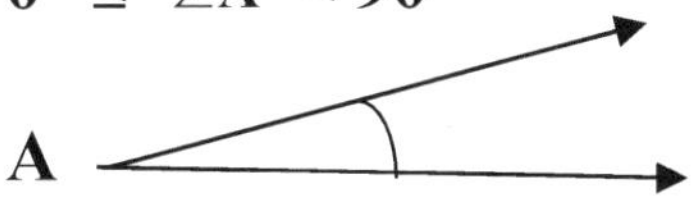

Figure 4

An angle larger than 90^0 but smaller than 180^0 called an obtuse angle:

$90^0 < \angle A < 180^0$ **$\angle A = 120^0$**

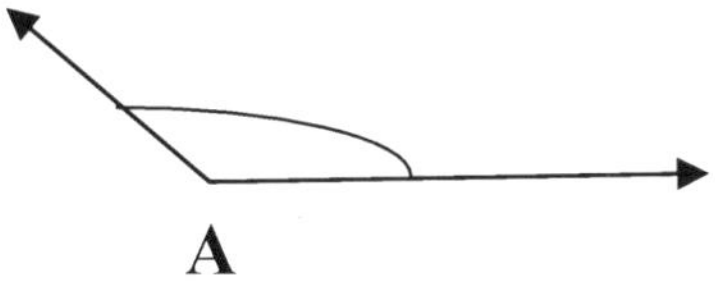

Figure 5

A straight angle is an angle that measures 180^0

A straight angle formed by straight lines.

$\angle A = 180^0$

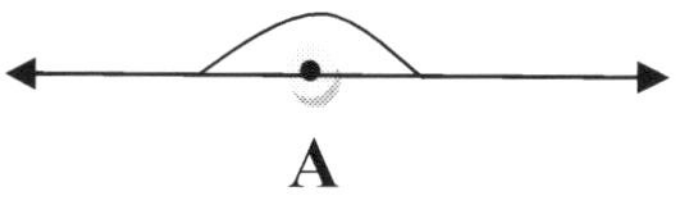

Figure 6

A right angle is half of a straight and equal 90^0

$\angle A = 90^0$

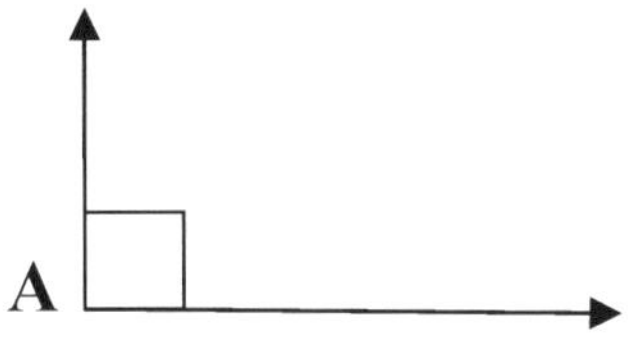

Figure 7

A circle is a full one revolution and equals 360^0

$\angle O = 360^0$

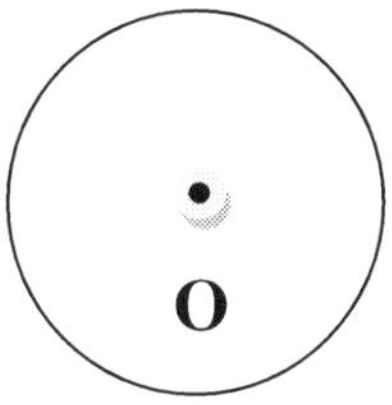

Figure 8

Grade 8 Chapter 3 Lesson 3

Pairs of Angles

Two angles are complementary if the sum of their measures is 90^0:

$\angle 1 + \angle 2 = 90^0$ $\qquad$ $30^0 + 60^0 = 90^0$ $\qquad$ $45^0 + 45^0 = 90^0$

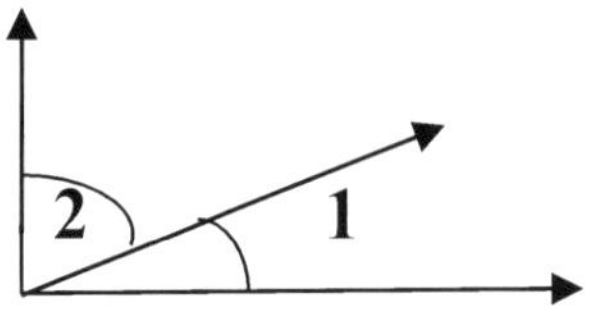

Figure 9

Two angles whose sum is equals 180^0 are supplementary angles.
Two angles with one common side and another two sides formed straight line called supplementary.

$\angle 1 + \angle 2 = 180^0$ $\qquad$ $60^0 + 120^0 = 180^0$

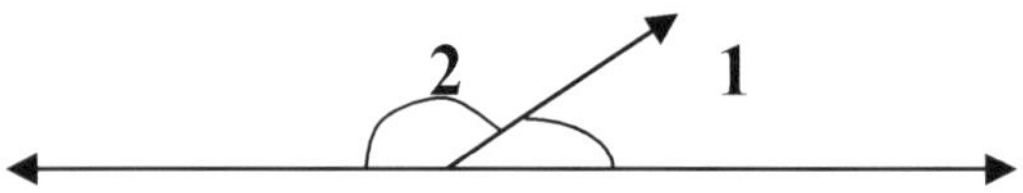

Figure 10

Two intersection lines formed four angles and two pair of verticals angles.
Vertical angles are congruent. $\angle 1 = \angle 3$ and $\angle 2 = \angle 4$ are verticals

$\angle 1 = \angle 3 = 30^0$

$\angle 2 = \angle 4 = 150^0$

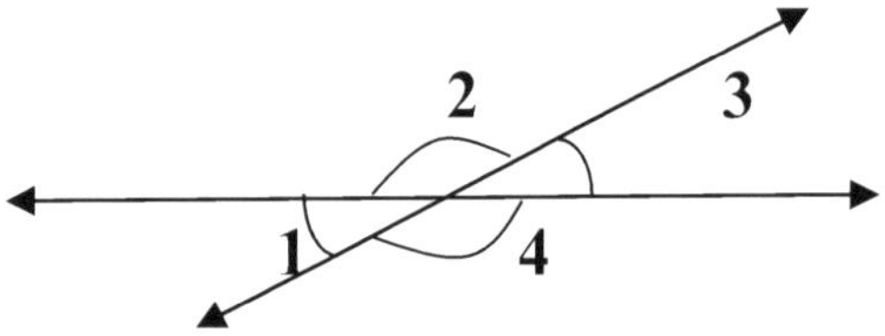

Figure 11

Two lines are perpendicular if the angle between them is 90^0.
Segments that lay on perpendicular lines called perpendicular segments:

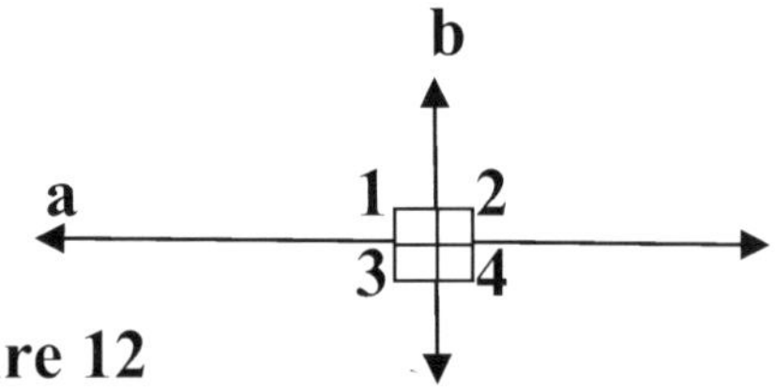

$\angle 1 = \angle 2 = \angle 3 = \angle 4 = 90^0$

$a \perp b$

Figure 12

Grade 8 Chapter 3 Lesson 3

When two parallel lines are crossed by a transversal, eight angles are formed.

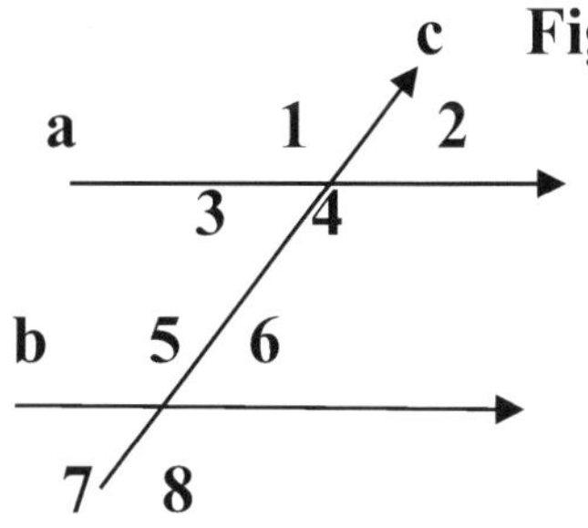

Figure 13 $a \parallel b$ **transversal c**

$\angle 2 = 30^0$ $\angle 1 = 180^0 - 30^0 = 150^0$

$\angle 3 \cong \angle 6$ $\angle 4 \cong \angle 5$ **(interior)**

$\angle 1 \cong \angle 8$ $\angle 2 \cong \angle 7$ **(exterior)**

$\angle 1 \cong \angle 5$ $\angle 2 \cong \angle 6$ **(corresponding)**

Problems with angles

Example 1

Second angle is 0.5 of the first angle of supplementary angles. Find both angles. The first angles is represented by x degree

The second angle represents 0.5 x.
The sum of the supplementary angle is 180^0.

Equation:

$x + 0.5\,x = 180$ $1.5\,x = 180$ $x = 180 \div 1.5 = 1800 \div 15 = 120$

$0.5 \cdot 120 = 60^0$ **First angle is 120^0. Second angle is 60^0**

Figure 14

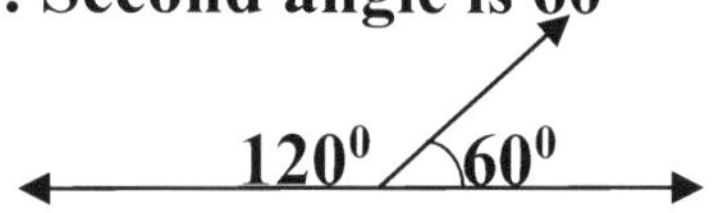

Grade 8 Chapter 3 Lesson 3
Problems with angles

Example 2

Supplementary angles have ratio like of 9 to 3. Find the angles.
Let's represent 1 part for x. The first angle is 9x, the second angle is 3x.

Equation: $9x + 3x = 180$ $12x = 180$ $x = 180 \div 12$ $x = 15^0$
First angle is $9 \cdot 15 = 135^0$ **Second angle is** $3 \cdot 15 = 45^0$

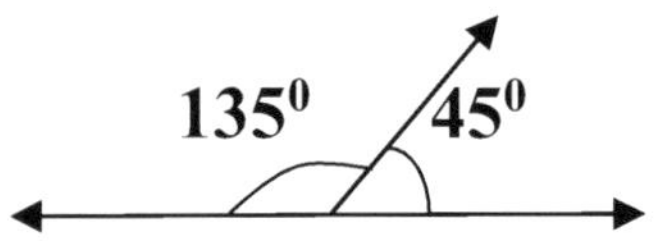

Figure 15

Example 3

Find the supplementary angles if the first angle is 30 more than the second angle.
Let's represent the second angle as x. The first angle is x + 30

Equation: $x + 30 + x = 180$ $2x + 30 = 180$ $2x + 30 - 30 = 180 - 30$
$2x = 150$ $x = 150 \div 2$ $x = 75^0$ $75^0 + 30^0 = 105^0$

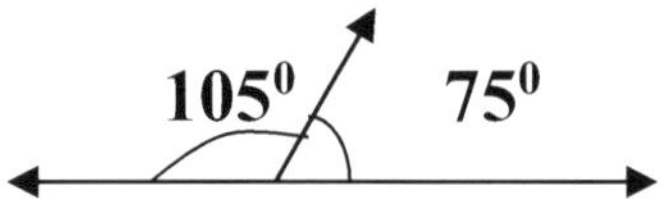

Figure 16

Example 4

Find the supplementary angles if the second angle is 30 less than the first angle. Find angles.
The first angle is x. Second angle is x – 30. Equation: $x + x - 30 = 180$

$2x - 30 = 180$ $2x - 30 + 30 = 180 + 30$ $2x = 210$ $x = 210 \div 2$
$x = 105^0$ $105 - 30 = 75^0$

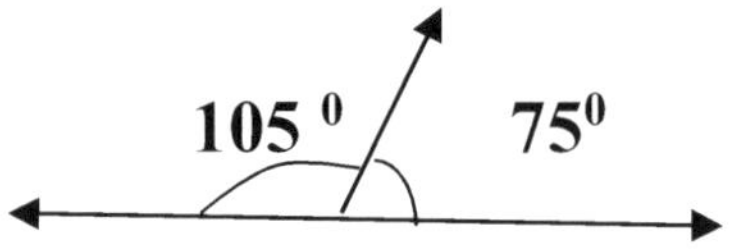

Figure 17

Grade 8 Chapter 3 Lesson 3

Angles
Review 1 A
Draw the pictures
Solve the problems

1) Given: $\angle 2 = 120^0$ $\angle 1 = 45^0$ Find: $\angle 2$ - $\angle 1$

2) Given: $\angle 2 = 90^0$ $\angle 1 = 45^0$ Find: $\angle 1 + \angle 2$

3) Given: $\angle 2 = 150^0$ $\angle 1 = 60^0$ Find: $\angle 2 - \angle 1$

4) Given: $\angle 1 = 30^0$ Find: $5\angle 1$

5) Given: $\angle 2 = 135^0$ Find: $\angle 2 \div 3$

6) Given: $\angle 2 = 120^0$ Find: $\angle 2 \div 2$

7) Find one of the complementary angles if another angle equals 60^0

8) Find one of the supplementary angles if another angle equals 60^0

9) Find one of the supplementary angles if another angle equals 150^0

10) Find one of the supplementary angles if another angle equals 135^0

Grade 8 Chapter 3 Lesson 3
Review 1 B

Solve the problems
Draw the pictures

1) One angle is 0.2 of the second angle of supplementary angles. Find both angles.

2) One angle is 0.8 of the second angle of supplementary angles. Find both angles.

3) Supplementary angles have a ratio of 9 to 11. Find the angles.

4) Supplementary angles have a ratio of 7 to 8. Find the angles.

5) Find the second angle, which is 9 times more than the first angle of supplementary angles.

6) One of the supplementary angles is 5 times less than the second angle. Find the biggest angle.

7) Find the supplementary angles if one angle is 20 more than the other angle.

8) Find the supplementary angles if one angle is 40 less than the other angle.

9) One of the supplementary angles is 4 time less than the second angle. Find the biggest angle.

10) Supplementary angles have a ratio of 4 to 5. Find the angles.

Grade 8 Chapter 3 Lesson 4

Triangles

A triangle is a geometric figure that contains three points and three segments connecting its points.

ΔABC Triangle has 3 vertexes, 3 sides, 3 angles.

The three points of a triangle has vertexes A, B, C.

The three segments of the triangle have sides called AB, BC, AC or a, b, c.

The three angles formed by the sides of the triangle called the inner angles of triangles: $\angle A, \angle B, \angle C$

Each side of a triangle is less the sum of the other two sides:

$a < b + c$	$b < a + c$	$c < a + b$
$6 < 12+8$	$12 < 6 + 8$	$8 < 6 + 12$

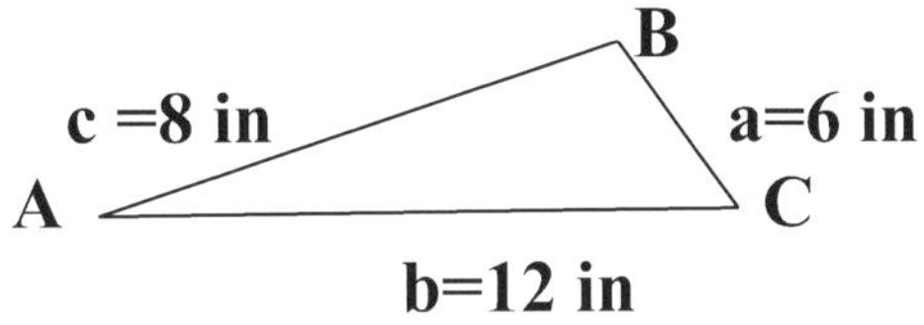

Figure 1

If three sides and three inner angles for one triangle accordingly equal three sides and three angles for another triangle then two triangles are congruent.

Figure 2

Grade 8 Chapter 3 Lesson 4

Different kind of triangles

An acute triangle contains three acute angles (Figure 3a)

$0^0 < A < 90^0$

$0^0 < B < 90^0$

$0^0 < C < 90^0$

A right triangle contains one right angle and two acute angles (Figure 3b) $C = 90^0$

An obtuse triangle contains one obtuse angle and two acute angles (Figure 3c)
$90^0 < C < 180^0$

Sum of inner angles of a triangles equals 180^0 $A + B + C = 180^0$

Example $50^0 + 60^0 + 70^0 = 180^0$

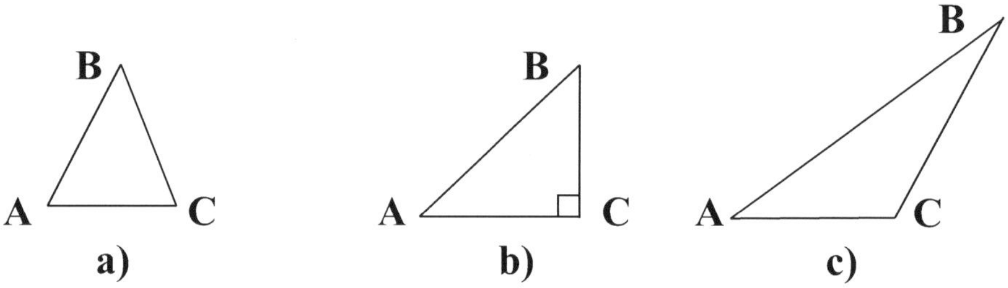

Figure 3

Grade 8 Chapter 3 Lesson 4

Different elements of triangles

A triangle has vertexes, sides, angles and bisectors, medians, and heights. A ray starting from the vertex of the angle, which divides this angle by two equal parts, is called the bisector and denotes OB = l.

The segment of the bisector of the angle of the triangle connecting the vertex of the triangle with a point on the opposite side called the bisector of the triangle.

A triangle has three bisectors, which intersect at one point inside the triangle:

$\Delta ABC \quad AB = c \quad AC = b \quad BC = a$

$\angle 1 = \angle 2 \quad \angle 3 = \angle 4 \quad \angle 5 = \angle 6 \quad AN = l_a \quad BD = l_b \quad CM = l_c$

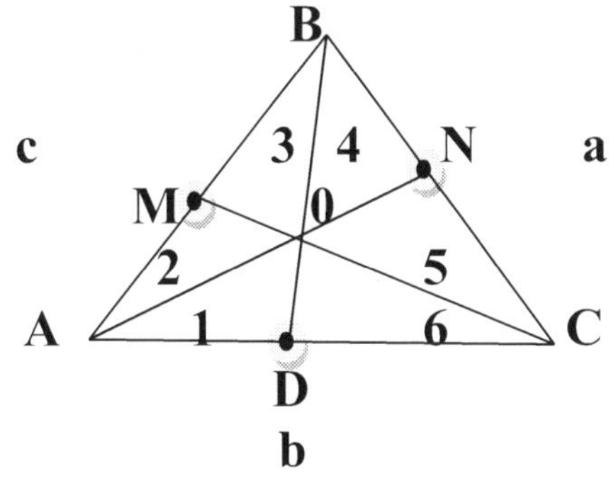

Figure 4

A segment, connecting the vertex to the middle of the opposite side called the median of the triangle.

A triangle has 3 medians, which intersect at one point inside the triangle.

$\Delta ABC \quad AM = MB \quad BN = NC \quad AD = DC \quad AN = m_a \quad BD = m_b \quad CM = m_c$

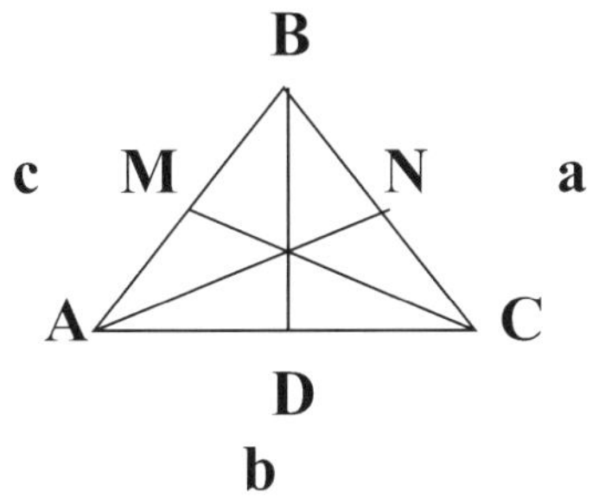

Figure 5

Grade 8 Chapter 3 Lesson 4
Height of triangle

Perpendicular from vertex of triangle to straight line contains opposite side, called height of triangle Δ ABC

A triangle has 3 heights, which intersect in one point.

In an acute triangle, the 3 heights intersect inside of the triangle Point O: (Figure 6a)

In a right triangle, the 3 heights intersect at the vertex of the right angle Point C: (Figure 6b) CM $\perp$ AB

In an obtuse triangle, the 3 heights intersect outside of the triangle – Point O: (Figure 6c)

$\Delta ABC \quad AN \perp BC \quad BD \perp AC \quad CM \perp AB \quad AN = h_a \; BD = h_b \; CM = h_c$

(Figure 6 a and Figure 6 c)

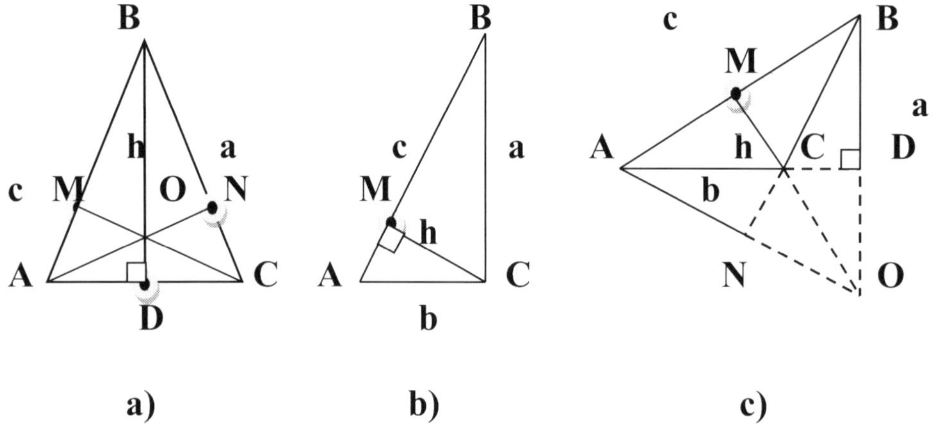

Figure 6

Grade 8 Chapter 3 Lesson 4

Special triangles

The sum of the three lengths of the sides of a triangle called the perimeter.

A scalene triangle has three different sides: P = a + b + c

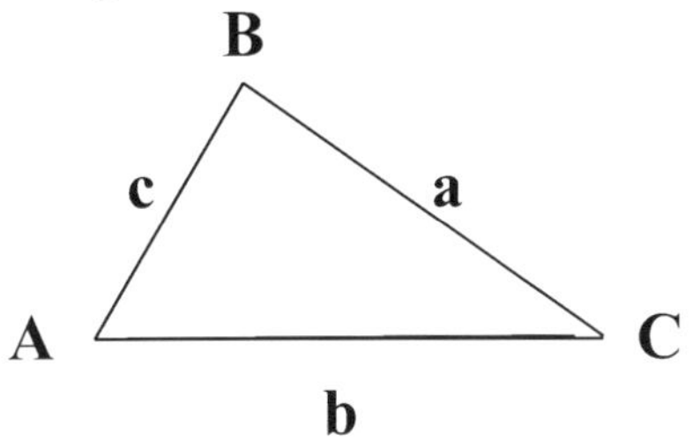

Figure 7

An isosceles triangle is a triangle with two equal sides.
The two equal sides are called legs a and a, the third side is called base b:

If AB = BC = a and AC = b than P = 2a + b
Angles on the base isosceles triangle are congruent

If in $\triangle ABC$ AB = BC than $\angle A = \angle C$
In an isosceles triangle, the bisector from the vertex to the base is the median and height.

.

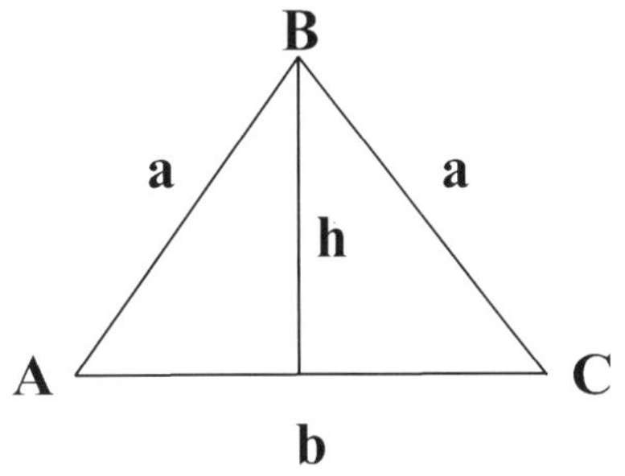

Figure 8

Grade 8 Chapter 3 Lesson 4

Equilateral Triangle

An equilateral triangle is a triangle with three equal sides:

If AB = BC = AC = b то P = 3b.

If in $\triangle ABC$ AB = BC = AC then $\angle A = \angle B = \angle C$

If in $\triangle ABC$ AB = BC = AC and $\angle A + \angle B + \angle C = 180^0$

$\angle A = \angle B = \angle C = 180^0 \div 3 = 60^0$

In an equilateral triangle, all three bisectors are medians and heights:

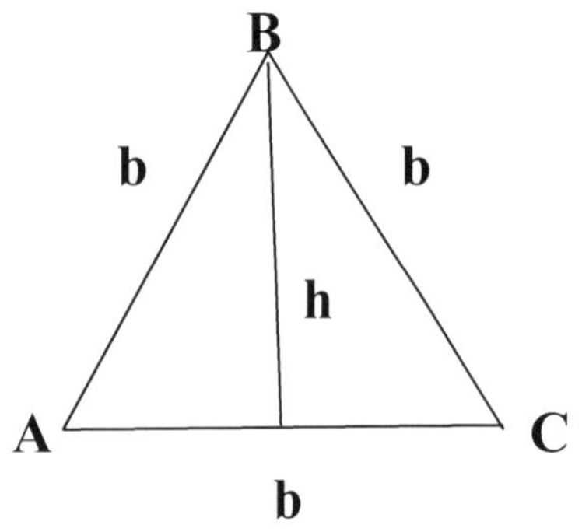

Figure 9

Grade 8 Chapter 3 Lesson 4
Triangles
Review 1 A
Solve the problems
Draw the pictures

1) **Find the third angle of a triangle if the two other angles are 40^0 and 50^0.**
2) **Find the third angle of a triangle if the two other angles are 55^0 and 60^0.**

3) **Find the angle of a right triangle if one of the angles is 45^0.**

4) **Find the angle of a right triangle if one of the angle is 30^0.**

5) **Find the perimeter of a triangle if its sides equal 5, 6, and 7 cm.**

6) **Find the third side of a triangle if two of its sides are 8 m and 9m, and the perimeter is 27 m.**

7) **Find the perimeter of an isosceles triangle if the base is 16 in and one of legs is 10 in**

8) **Find the legs of an isosceles triangles if the base is 8 ft and the perimeter is 18 ft**

9) **Find the base of an isosceles triangle if one the leg is 10 yd and perimeter is 36 yd**

10) **Find the perimeter of an equilateral triangle if its side is 6 ft.**

Grade 8 Chapter 3 Lesson 4
Review 1 B
Solve the problems
Draw the pictures

1) Find the side of an equilateral triangle if its perimeter is 24 in.

2) Find the perimeter of an isosceles triangle if one of legs is 20 cm and the base is 10 cm

3) Find one leg of an isosceles triangle if its base is 1.3 in and its perimeter is 2.5 in.

4) The first side of a triangle is 17 cm of a triangle. The second side of the triangle is double of the first side. Third side of the triangle less the second side of the triangle on 10 cm. Find the perimeter of the triangle.

5) Find the base of an isosceles triangle if its perimeter 3.4 ft and it's one of legs is 1.3 ft.

6) The perimeter of an isosceles triangle is 40 cm. The leg is twice than the base of the triangle. Find the sides of the triangle.

7) The perimeter of an isosceles triangle is 26 in. Its base is 6 times less than its leg. Find the sides of the triangle.

8) The perimeter of an isosceles triangle is 20 in. Its leg is 2 in less than its base. Find the sides of the triangle.

9) The perimeter of an isosceles triangle is 40 in. Its leg is 5 in more than its base. Find the sides of the triangle.

10) The perimeter of an isosceles triangle is 22 cm. Its base is 2 cm less than its leg. Find the sides of the triangle.

Grade 8 Chapter 3 Lesson 5
Right triangles

A right triangle contains one right angle and two acute angles: $C = 90^0$
For example, special right triangles: 30^0, 60^0, 90^0 and 45^0, 45^0, 90^0

The sides of a right triangle that form the right angle are called legs a and b
The side of a right triangle that is opposite the right angle is called hypotenuse (c)

The perimeter of a triangle is equal to the sum of its sides.
If hypotenuse AB = c, legs AC = b and BC = a then P = a + b + c

In a right triangle, the hypotenuse is more than the any other legs

$C = 90^0$	$A < 90^0$	$B < 90^0$
$A < C$	$a < c$	$c > a$
$B < C$	$b < c$	$c > b$

The sum of two acute angles in a right triangle is equal to 90^0
$A + B + C = 180^0$ $\quad$ $C = 90^0$ $\quad$ $A + B = 90^0$

The property for the sides of a right triangle called the Pythagorean Theorem.
In a right triangle, the square of the hypotenuse equals the sum of the square of the legs.
$c^2 = a^2 + b^2$ or $a^2 + b^2 = c^2$ $\quad$ $a^2 = c^2 - b^2$ or $b^2 = c^2 - a^2$

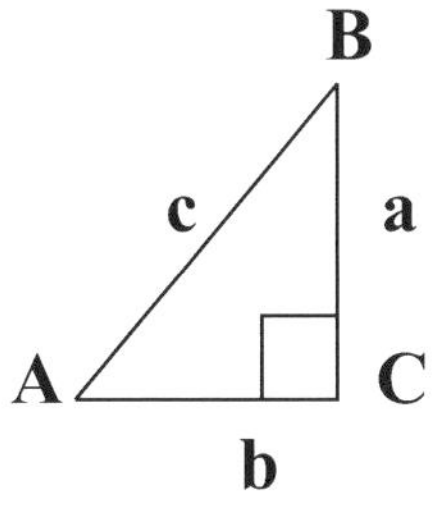

Figure 1

Grade 8 Chapter 3 Lesson 5

Pythagorean Theorem

Triangle with sides 3, 4, 5 people knew before proven Pythagorean theorem

$3^2 + 4^2 = 5^2$ $\quad$ $9 + 16 = 25$

Right triangles, in which sides expressed in natural numbers, called Pythagorean triples $\quad$ $6^2 + 8^2 = 10^2$ $\quad$ $36 + 64 = 100$

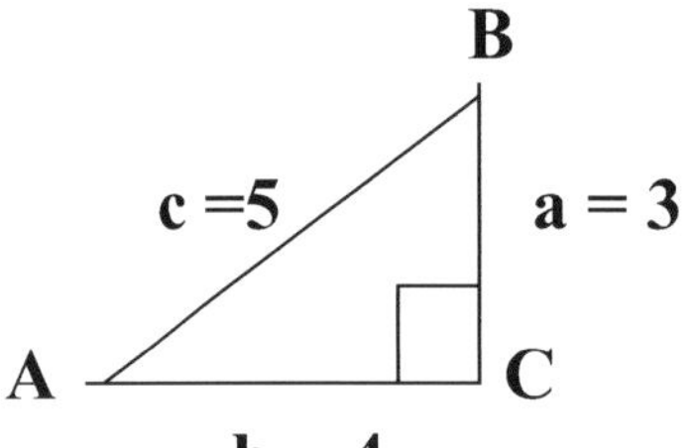

Figure 2

Examples for Right Triangle ABC (Figure 2)

a) $c = 17$ ft $\quad$ $b = 8$ ft $\quad$ $a = ?$ $\quad$ $a^2 + b^2 = c^2$

$a^2 + 8^2 = 17^2$ $\quad$ $a^2 + 64 = 289$ $\quad$ $b^2 = 289 - 64$ $\quad$ $b^2 = 225$

$b = 15$ft

b) $a = 20$ yd $\quad$ $c = 29$ yd $\quad$ $b = ?$ $\quad$ $c^2 = a^2 + b^2$

$29^2 = 20^2 + b^2$ $\quad$ $841 = 400 + b^2$ $\quad$ $441 = b^2$ $\quad$ $b = 21$ yd

c) $a = 52$ m $\quad$ $b = 39$ m $\quad$ $c^2 = a^2 + b^2$ $\quad$ $c^2 = 52^2 + 39^2$

$c^2 = 2704 + 1521$ $\quad$ $c^2 = 4225$ $\quad$ $c = 65$ m

Rectangle

d) **Find the length of a rectangle, if the diagonal is 15 in and the width is 9 in.**

$d = 15$ in $\quad$ $w = 9$ in $\quad$ $l^2 + w^2 = d^2$ $\quad$ $l^2 + 9^2 = 15^2$

$x^2 + 81 = 225$ $\quad$ $x^2 = 225 - 81$ $\quad$ $x^2 = 144$ $\quad$ $x = 12$ in

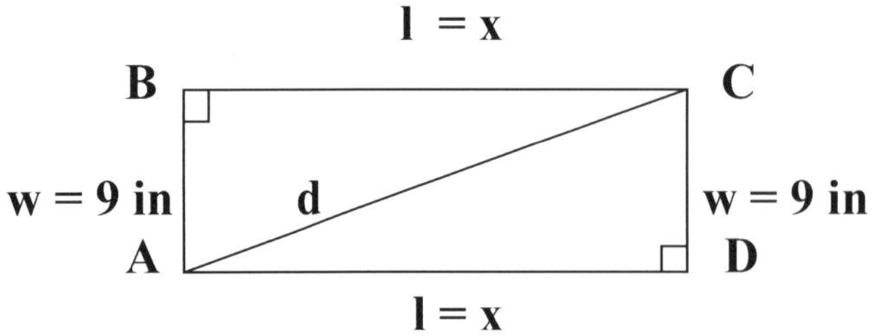

Figure 3

Grade 8 Chapter 3 Lesson 5

Right triangle

Review 1 A

Draw the right triangles
Find the missing side of each right triangle
$a^2 + b^2 = c^2$

	a	b	c		a	b	c
1)	?	4	5	2)	6	?	10
3)	9	12	?	4)	?	16	20
5)	15	?	25	6)	18	24	?
7)	?	28	35	8)	24	?	40
9)	27	36	?	10)	?	40	50
11)	33	?	55	12)	36	48	?
13)	?	52	65	14)	42	?	70
15)	45	60	?	16)	?	64	80
17)	51	?	85	18)	54	72	?
19)	?	76	95	20)	60	?	100
21)	5	12	?	22)	?	24	26
23)	15	?	39	24)	20	48	?
25)	?	60	65				

Grade 8 Chapter 3 Lesson 5

Review 1 B

Draw the right triangles

Find the missing side of each right triangle

$$a^2 + b^2 = c^2$$

	a	b	c		a	b	c
1)	3	?	7.8	2)	0.7	2.4	?
3)	?	4.8	5	4)	2.1	?	7.5
5)	2.8	9.6	?	6)	?	2.1	2.9
7)	4	?	5.8	8)	0.8	1.5	?
9)	?	3	3.4	10)	2.4	?	5.1
11)	0.9	4	?	12)	?	8	8.2
13)	1.1	?	6.1	14)	1.3	8.4	?
15)	?	6.3	8.7	16)	3	?	$7\frac{4}{5}$
17)	$1\frac{2}{5}$	4.8	?	18)	?	$7\frac{1}{5}$	$7\frac{1}{2}$
19)	$2\frac{4}{5}$	?	10	20)	4	$4\frac{1}{5}$	?
21)	?	$1\frac{1}{2}$	1.7	22)	$1\frac{3}{5}$	?	3.4
23)	$2\frac{2}{5}$	4.5	?	24)	?	8	$8\frac{1}{5}$
25)	1.3	?	$8\frac{1}{2}$				

Grade 8 Chapter 3 Lesson 6

Perimeter

A polygon means a many-sided figure.
A polygon has three or more sides of the figure.

Points called vertexes and segments called sides.
A triangle has 3 vertexes, 3 sides, and 3 inner angles (Figure 1a)

A quadrilateral has 4 vertexes, 4 sides, and 4 inner angles (Figure 1b)
A pentagon has 5 vertexes, 5 sides, and 5 inner angles (Figure 1c)

A hexagon has 6 vertexes, 6 sides, and 6 angles (Figure 1d)
Two vertexes of polygon, which belong to one side, called adjacent.

A segment connecting any two nonadjacent vertexes called diagonal (d) Figure 1 b

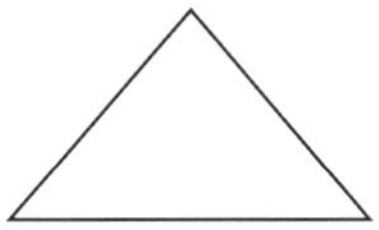

Figure 1 a

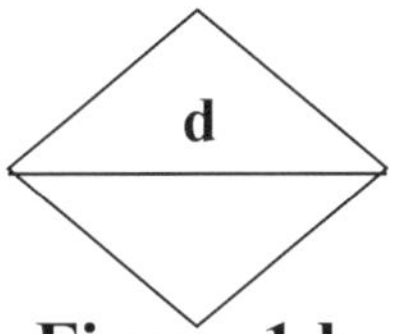

Figure 1 b

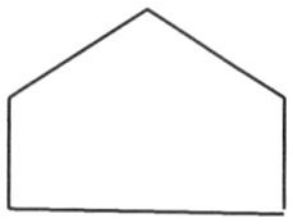

Figure 1 c

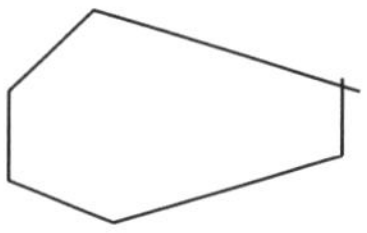

Figure 1 d

Quadrilateral

A quadrilateral has 4 vertexes, 4 sides, and 4 inner angles.
Two nonadjacent vertexes or sides of a quadrilateral are called opposite vertexes or sides.

The perimeter is the distance around a figure or the sum of the length of all sides
The measurement of the perimeter is in linear units:
1 in, 1 ft, 1 yd, 1 mi, 1 mm, 1cm, 1dm, 1m, 1 km

Grade 8 Chapter 3 Lesson 6

Parallelogram

A parallelogram is a quadrilateral in which the opposite sides are parallel.
If ABCD is a parallelogram, then AD || BC and AB || DC

Figure 2

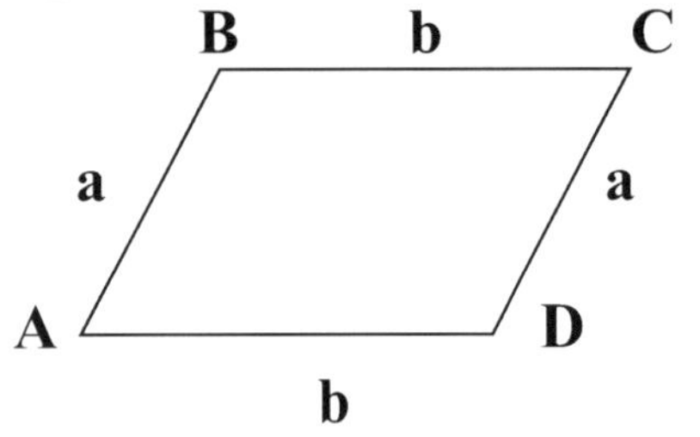

The sum of inner angles for any parallelogram is equal 360^0

$\angle A + \angle B + \angle C + \angle D = 360^0$

The sum of angles adjacent to one side of a parallelogram is equal to 180^0. Figure 2

If AB || DC then $\angle A + \angle D = 180^0$ **and** $\angle B + \angle C = 180^0$

If AD || BC then $\angle A + \angle B = 180^0$ **and** $\angle C + \angle D = 180^0$

Opposite angles of a parallelogram are equal:

If ABCD is a parallelogram, then $\angle A = \angle C$ **and** $\angle B = \angle D$

The opposite sides of a parallelogram are equal:

If ABCD is a parallelogram, then AB = CD and AD = BC

Parallelogram ABCD, where AB = a AD = b

The perimeter of a parallelogram with sides a and b can use any of the following formulas:

$P_{par} = a + b + a + b$ $\quad$ $P_{par} = 2a + 2b$ $\quad$ $P_{par} = 2(a + b)$

Rhombus

A rhombus is a parallelogram in which measures of the four sides are equal.

Figure 3

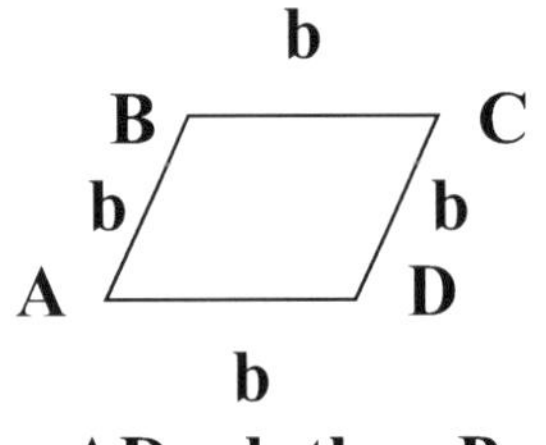

If AB = BC = CD = AD = b then P = b + b + b + b $\quad$ **P = 4b**

All properties of parallelogram are correct for a rhombus:

AB || DC $\quad$ **AD || BC** $\quad$ $\angle A = \angle C$ **and** $\angle B = \angle D$

Grade 8 Chapter 3 Lesson 6

Rectangle

A rectangle is parallelogram in which all four angles are right angles.

Figure 4

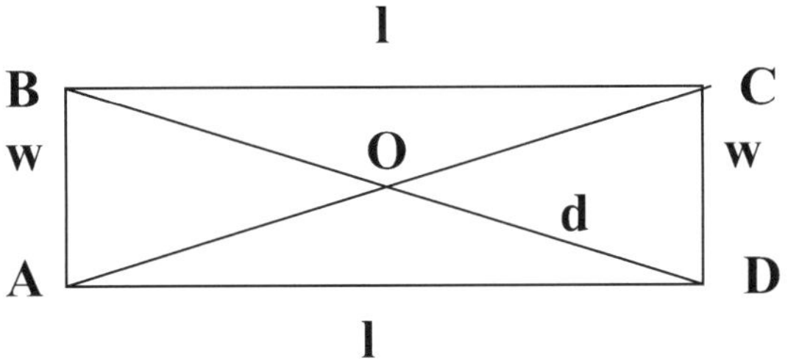

If ABCD is a rectangle, then $\angle A = \angle B = \angle C = \angle D = 90^0$

The dimensions of a rectangle are length l and width w.
The perimeter of a rectangle is equal to the sum of its sides.
If ABCD is a rectangle   AD = BC = l   AB = CD = w
P = l + w + l + w   P = 2l + 2w   P = 2(l + w)

All properties of a parallelogram are the same for a rectangle
A rectangle has its own property: the diagonals of a rectangle are congruent.
If ABCD is a rectangle, then AC = BD = d

Square

A square is a rectangle in which measures of all four sides are equal.
A square is a rhombus in which all four angles are right angles.
A square is a parallelogram in which all four sides are equal and all angles are right angles.

Figure 5

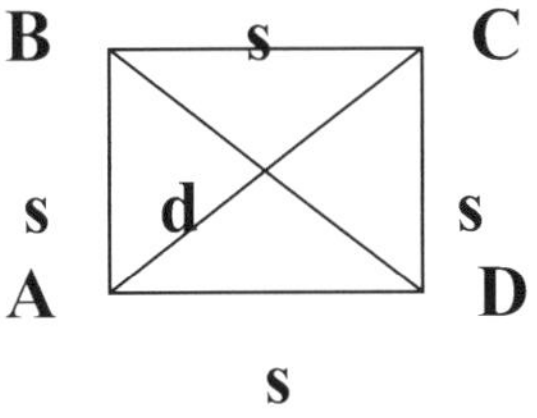

All properties for a rectangle, rhombus, and parallelogram are the same for a square. The perimeter of a square is equal to the sum of the four sides:
If AB = BC = CD = AD = s   P = s + s + s + s   P = 4s
$\angle A = \angle B = \angle C = \angle D = 90^0$ AC = BD = d

Grade 8 Chapter 3 Lesson 6

Trapezoid

A trapezoid is a quadrilateral that has exactly two opposite sides that are parallel. The two parallel sides called bases and the two non-parallel sides called legs.

Figure 6

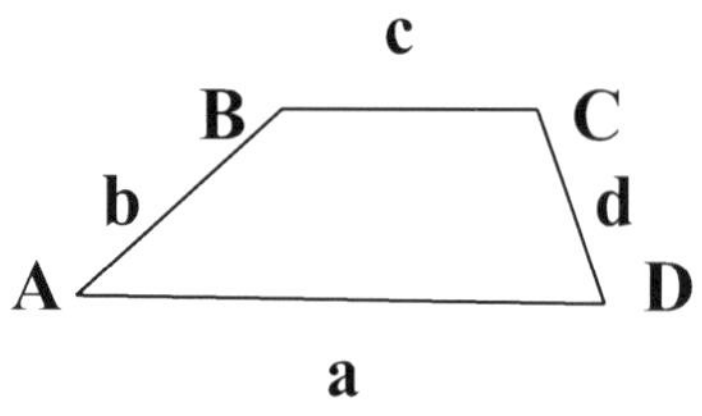

AD || BC AD and BC are bases AB and CD are legs
If ABCD is a trapezoid, AD = a AB = b BC = c CD = d
P = a + b + c + d

A trapezoid called right trapezoid if it has one right angle:
If ABCD is a right trapezoid and $\angle A = 90^0$ then $\angle B = 90^0$

Figure 7

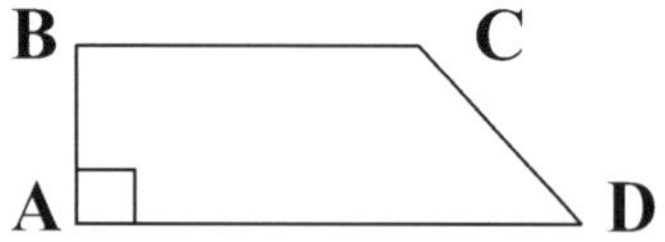

A trapezoid is called an isosceles trapezoid if it has measures of legs are equals.
If ABCD is an isosceles trapezoid and AD || BC than AB = CD

Figure 8

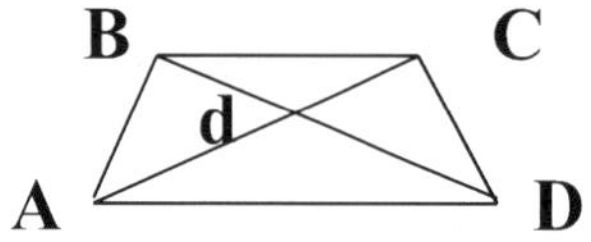

An isosceles trapezoid has diagonals whose length are equal.
AC = BD = d

Grade 8 Chapter 3 Lesson 6

Kite

A kite is quadrilateral in which there are two pairs of equal adjacent sides.
If ABCD is a kite, then AB = AD and CB = CD

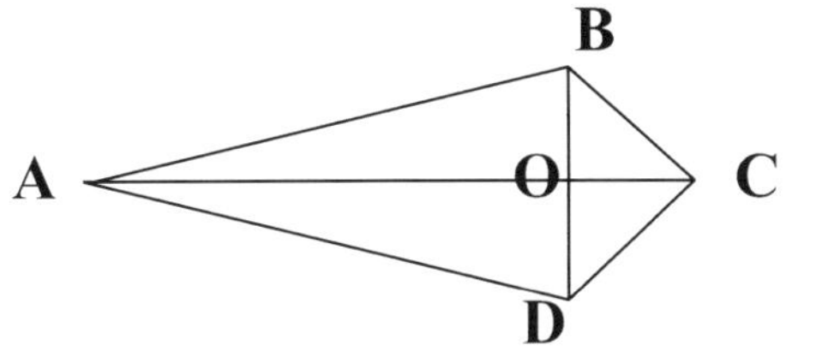

Figure 9

Regular polygons

A polygon called regular which measures of all sides and all angles are equal.
A regular triangle is equilateral triangle and its perimeter: P = 3 s.
A regular quadrilateral is the square and its perimeter: P = 4 s
A regular pentagon is a polygon that measures of five equal sides: P = 5 s. (Figure 10a)
A regular hexagon is a polygon that measures of six equal sides: P = 6 s (Figure 10 b)

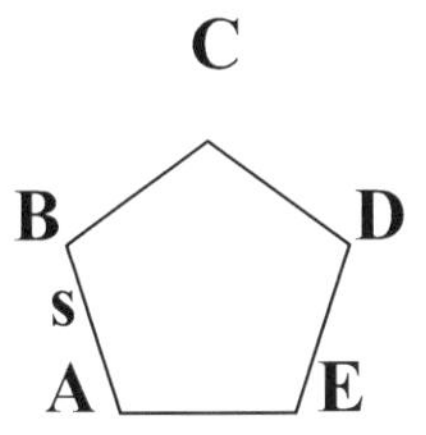

Figure 10 a

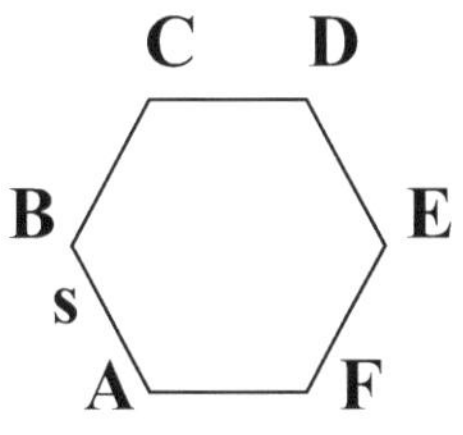

Figure 10 b

Formulas

The perimeter of a triangle is the sum of the length of all three sides.
P = a + b + c
The perimeter of a square is four times the length of the sides. P = 4 s
The perimeter of a rhombus is four times the length of the sides. P = 4 b
The perimeter of a rectangle is the sum of twice of the length and twice of the width: P = 2 l + 2 w
The perimeter of a parallelogram is the sum of twice of the adjacent sides.
P = 2 a + 2 b

Grade 8 Chapter 3 Lesson 6

Perimeter
Review 1 A
Solve the problems
Draw the pictures

1) Find the perimeter of a triangle, if its sides equal 6cm, 8cm, 10 cm.
2) Find the perimeter of an equilateral triangle, if a side equals 5 in.
3) Find a side of an equilateral triangle, if the perimeter equals 30 ft.
4) Find the perimeter of an isosceles triangle, if its sides equal 10cm, 10 cm, and 8cm.
5) Find the perimeter of a square, if its side equal 5 yd.

6) Find the side of a square, if its perimeter equals 40 in.
7) Find the perimeter of a rhombus, if its sides equal 10 dm.
8) Find the side of a rhombus, if its perimeter equals 24 cm.
9) Find the perimeter of a rectangle, if its length equals 6 cm, and the width is 4 cm
10) Find the length of a rectangle if its perimeter is 44 in, and its width is 10 in.

11) Find the width of a rectangle if its perimeter is 88 ft and its length is 30 ft
12) Find the perimeter of a parallelogram, if the length of its sides are 10m and 8m
13) Find the side of a parallelogram if its perimeter is 60 ft and one of its side 10 ft
14) Find the perimeter of a trapezoid, if its sides are 10cm, 8cm, 4cm, and 6cm
15) Find the perimeter of regular pentagon, if each side equals 10 ft.

16) Find the side of a regular pentagon if its perimeter is 50 yd.
17) Find the perimeter of a hexagon, if its side equal 4m, 6m, 4m, 2m, 6m, and 2m.
18) Find the perimeter of a regular hexagon, if each side equals 5 ft.
19) Find the side of a regular hexagon if its perimeter is 60 m.
20) Find perimeter of a kite, if its sides are equals 6 in and 10 in.

Grade 8 Chapter 3 Lesson 7

Area

The area is the number of square units that covers a surface.
Congruent polygons have the same areas: if $F_1 = F_2$ then $A_1 = A_2$
If a polygon formed by two different figure, then the area of the polygon equals the sum of the areas for each figure:
if $F = F_1 + F_2$ then $A = A_1 + A_2$
Measurement Area is Square Unit
1 mm², 1 cm², 1 dm², 1 m², 1 km² and 1 in², 1 ft², 1 yd², 1 mi²

1 cm = 10 mm	**1 ft = 12 in**	**1 yd = 3 ft**
1 sq.cm = 100 sq.mm	**$1\ ft^2 = 144\ in^2$**	**$1\ yd^2 = 9\ ft^2$**

Square

The area of a square is equal to the square of its sides:
ABCD is square $A = s \cdot s$ $\quad A_{\square} = s^2$
Example: s = 6 in. $A_{\square} = ?$ $\quad A_{\square} = s^2 = 6^2 = 6 \cdot 6 = 36\ in^2$

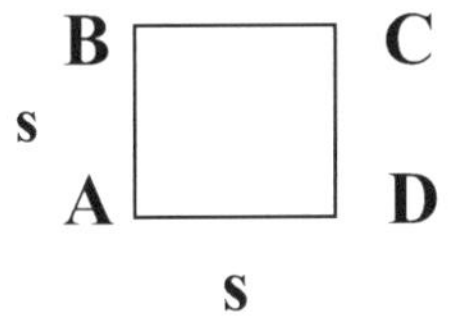

Figure 1

Rectangle

The area of a rectangle is equal to the product, length by its width.
If ABCD is a rectangle, AD = l, AB = w then $A_{rec} = l \cdot w$
Example: l = 10 m $\quad$ w = 5 m $\quad A_{rec} = l \cdot w = 10\ m \cdot 5\ m = 50\ m^2$

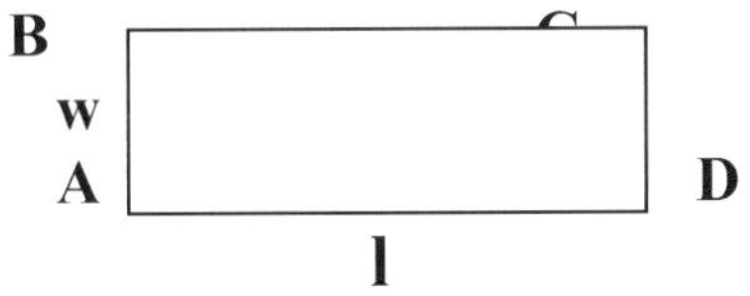

Figure 2

Rectangle and Square

Find the area of a figure that contains a rectangle and a square together.
Example: ABCD a rectangle, DCMN a square
if l = 10 in, w = 5 in, s = 5 in. $A = A_1 + A_2 = 10 \cdot 5 + 5^2 = 50 + 25 = 75\ in^2$

Figure 3

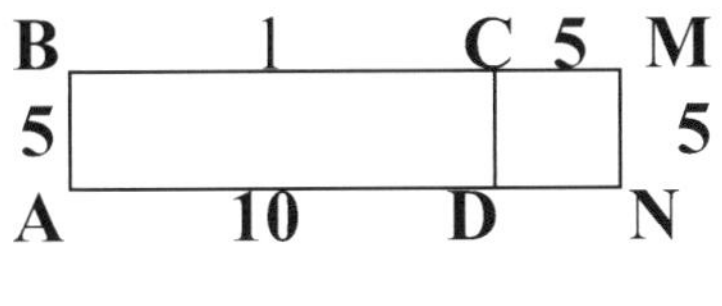

Grade 8 Chapter 3 Lesson 7

Rhombus

The area of a rhombus is equal to the product, base by its height.
If ABCD is a rhombus, AD = b, BM = h then $A_{rh} = b \cdot h$

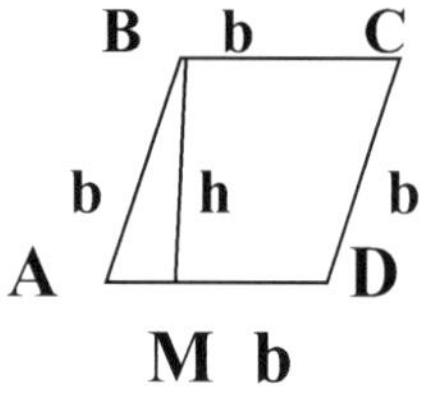

Example: b = 10 cm h = 5 cm

$A_{rh} = bh$ $A = 10 \cdot 5 = 50\ cm^2$

Figure 4

Parallelogram

One sides of a parallelogram called the base.
Perpendicular from any point on the opposite side to the line containing the base called the height of the parallelogram.
The area of a parallelogram is equal to the product, base by its height:
If ABCD is a parallelogram, AD = b BM = h then $A_{par} = b\ h$

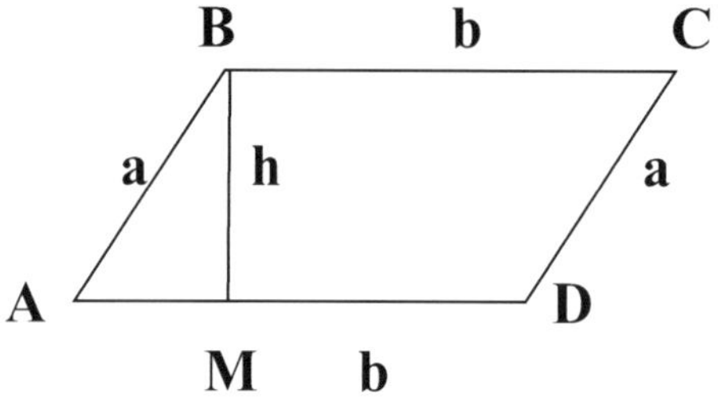

Example:

b = 20 m h = 10 m

$A_{par} = b\ h = 20 \cdot 10 = 200\ m^2$

Figure 5

Trapezoid

Perpendicular from any point of the upper base to the line containing the l ower base called the height of the trapezoid.
The area of a trapezoid is equal product half of height by sum of bases
If ABCD is a trapezoid, AD = a BC = c BM = h then $A_{tr} = 0.5h\ (a + c)$

Figure 6

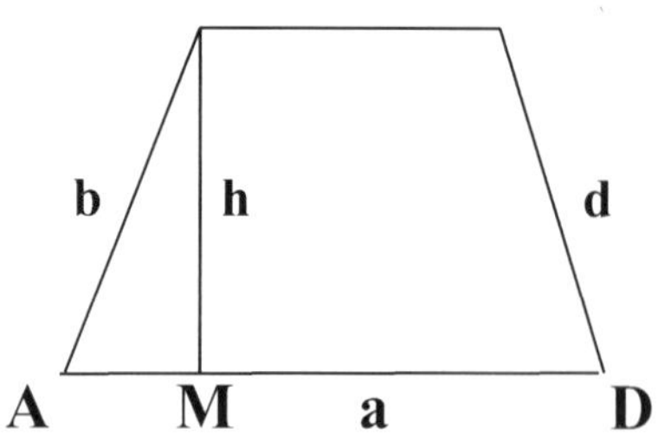

Example:

a = 12 m c = 8 m h = 6 m

$A_{tr} = 0.5\ h\ (a + c)$

$A_{tr} = 0.5 \cdot 6\ (12 + 8) = 3 \cdot 20 = 60\ m^2$

Grade 8 Chapter 3 Lesson 7

Formula for the Area of a Triangle

One of sides of a triangle called the base.
Perpendicular from any vertex to the line containing the base, is called the height of the triangle.
The area of a triangle equal half of product, base by height.
If ΔABC $AC = b$ $BM = h$ then $A_\Delta = 0.5bh$

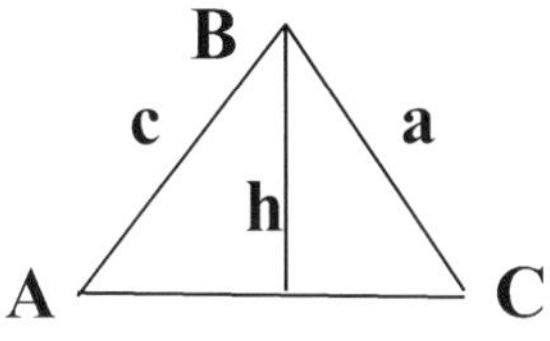

Example:
$b = 5$ cm $h = 4$ cm
$A_\Delta = 0.5\ bh = 0.5 \cdot 5 \cdot 4 = 0.5 \cdot 20 = 10\ cm^2$

Figure 7 b

Hero's Formula for Area of a Triangle

A triangle has sides a, b, c and a semi perimeter $s = (a + b + c) \div 2$
Area of triangle is equal $A_\Delta = \sqrt{s(s-a)(s-b)(s-c)}$

Figure 8 B

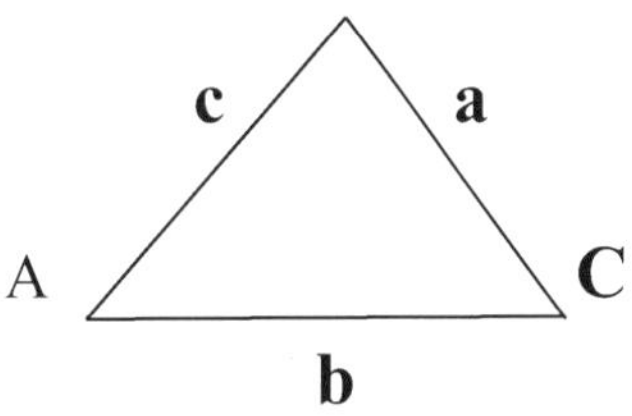

Example:

Find the area of a triangle with sides 3m, 4m, 5m.
$s = (3 + 4 + 5) \div 2 = 12 \div 2 = 6$

$A_\Delta = \sqrt{6(6-3)(6-4)(6-5)} = \sqrt{6 \cdot 3 \cdot 2 \cdot 1} = \sqrt{36} = 6\ м^2$

Right triangle

The area of a right triangle equal half of the product of its legs.
If ΔABC $C = 90^0$ $AC = b$ $BC = a$ than $A_\Delta = 0.5ab$
Example: $a = 3$ ft $b = 4$ ft $A_\Delta = 0.5\ ab = 0.5 \cdot 3 \cdot 4 = 6\ ft^2$

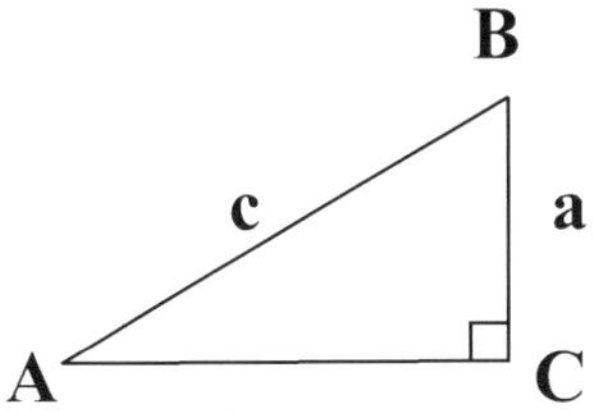

Figure 9 b

Grade 8 Chapter 3 Lesson 7
Area
Review 1 A
Solve the problems
Draw the pictures

1) Find the area of a square, if its side is 8 in.
2) Find the length of a square if the area of the square is 25 cm^2.

3) Find the area of a rectangle, if its dimensions are 9yd and 4 yd.
4) Find the length of a rectangle, if its area is 72 sq.cm, and its width is 8cm.

5) Find the width of a rectangle, if its area is 132 ft^2, and its length is 12ft.
6) Find the base of a rhombus if its height is 5m and its area is 60 m^2.

7) Find the height of a parallelogram if its base is 15in and its area is 150 in^2.
8) Find the height of a trapezoid, if its bases are 10m and 8m, and its area is 90 m^2.

9) Find the big base of trapezoid, if $c = 10$ in, $h = 6$ in, $A = 66$ in^2
10) Find the small base of a trapezoid, if $a = 10$ ft, $h = 4$ ft, $A = 36$ ft^2

11) Find the area of a triangle, if its base is 10m, and its height is 8m.
12) Find the base of a triangle, if its height is 6m, and its area is 60 m^2.

13) Find the area of a right triangle, if its legs are 10in and 8 in.
14) Find one of leg of a right triangle, if its other leg is 12m, and its area is 60 m^2.

15) Find the area of a triangle with sides 3, 4, and 5 м.
16) Find the area of a triangle with sides 6, 8, and 10in.
17) Find the area of a triangle with sides 5, 12, and 13ft.
18) Find the area of a triangle with sides 9, 12, and 15 cm.
19) Find the area of a triangle with sides 7, 24, and 25 yd.
20) Find the area of a triangle with sides 12, 16, and 20 dм.

Grade 8 Chapter 3 Lesson 8
Circle

A circle is a geometric figure, made up of all points, and ordered by a given distance from a given point.

The given point called the center of the circle.

A segment connecting the center with any point on the circle called the radius of the circle.

The given distance is a radius of the circle.

All radii are equal in one circle.

A segment passing through the center of the circle called the diameter of circle.

O – Center   AO = OB = r   AB is diameter   AB = d   (Figure 1)

The diameter of a circle is twice the radius: d = 2r

The degree of measurement of a full circle is 360 degrees.

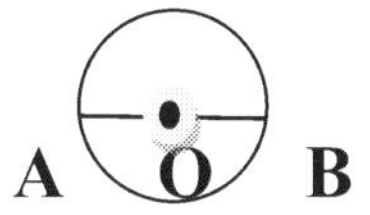

Figure 1

Grade 8 Chapter 3 Lesson 8
Arc and chord

Two points on a circle divide the circle into two parts, each part formed by arc of circle. The arc of a circle measured in degrees.

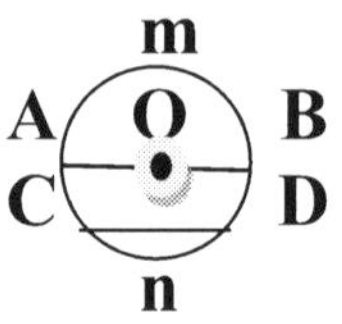

Figure 2

Circles can draw with compasses or patterns.

A segment connecting two points of a circle called a chord.

CD is a chord and ∪C n D is an arc (Figure 2)

The measure of two arcs together with the same ends is 360^0.
∪AmB + ∪A n B = 360^0

An arc called a semicircle if the segment connects the ends of the diameter. A semicircle is half of a circle. ∪AmB = ∪AnB = 180^0

Arcs can be compared, added, and subtracted: (Figure 3)

∪AB = ∪BC = ∪AC ∪AB < ∪ACB ∪ACB > ∪AB

∪AB + ∪BC = ∪ABC ∪ABC - ∪AB = ∪BC

Arc can be multiplied or divided by number:

∪AB · 2 = ∪ABC ∪A BC ÷ 2 = ∪AB

An angle with the vertex in the center of a circle called a central angle. Degree measured of a central angle of a circle is equal to the degree measured for an arc. ∠AOB = ∪A n B (Figure 4)

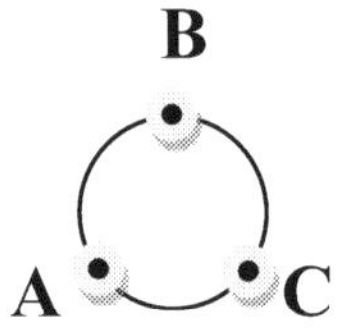

Figure 3

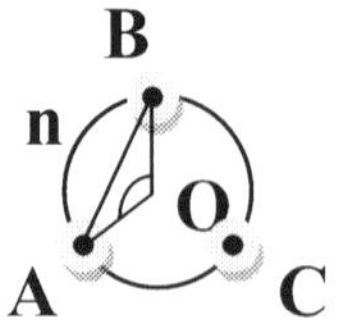

Figure 4

Grade 8 Chapter 3 Lesson 8
Circumference

The circumference is the distance around a circle.

Two formulas can used for finding the circumference.

$\frac{C}{d} = \pi$ $\quad d \times \frac{C}{d} = \pi \times d \qquad C = \pi d$ **where** $\pi \approx 3.14$ **or** $\pi \approx 3\frac{1}{7} \quad \pi \approx \frac{22}{7}$

First formula of circumference $C = \pi d$

Example 1: Find the circumference if d = 7 in, use $\frac{22}{7}$ for π.

$C = \pi d = \frac{22}{7} \cdot 7 \approx 22$ in

Second formula of circumference $C = 2\pi r$

If $d = 2r$ $\qquad C = \pi d \qquad C = \pi \cdot 2r \qquad C = 2\pi r$

Example 2: Find the circumference if r = 10 cm, use ≈ 3.14 for π

$C = 2\pi r = 2 \cdot 3.14 \cdot 10 \approx 62.8$ cm

Example 3: Find the exact answer of the circumference in terms of π

if r = 7 in. $\quad C = 2\pi r = 2\pi \cdot 7 = 2 \cdot 7\,\pi = 14\pi$ in

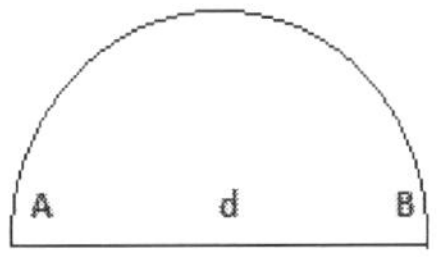

Circumference of a semicircle is
$C = 0.5\,(2\pi r) = \pi r$ or $C = 0.5\pi d$

Grade 8 Chapter 3 Lesson 8

Area of the Circle

The area of a circle is equal to the product of number π and the square of radius r: $A_c = \pi r^2$. The area of semicircle $A = 0.5\ \pi r^2$

Example 4: Find the area of the circle if radius r = 10 cm $\pi = 3.14$
$A_c = \pi r^2 = \pi \times 10^2 = 100\ \pi = 100 \times 3.14 \approx 314\ cm^2$

Example 5: Find the area of the circle if d = 14 cm $\pi \approx \frac{22}{7}$

d = 14 in r = 7 in $A_c = \pi r^2 = \frac{22}{7} \cdot 7^2 = \frac{22}{7} \times \frac{49}{1} = 22 \cdot 7 \approx 154\ in^2$

Example 6: Find the exact answer of the area in terms of π if r = 8 in.
$A = 0.5\ \pi r^2 = 0.5\pi \cdot 8^2 = 0.5 \cdot 64\ \pi = 32\pi\ in^2$

Concentric circles

Two circles with the same center but different radii called concentric circles. Concentrically circles form a ring.

The thickness of the ring is the equal the difference between its radii:
$h = R - r$

The area of a ring can be found by the formula:
$A_{ring} = \pi R^2 - \pi r^2 = \pi\ (R^2 - r^2)$

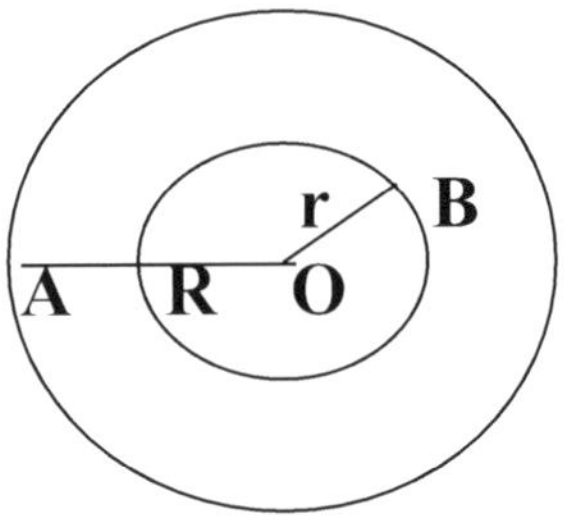

Example 7:
Find the area of a ring if its radii are
R = 10 in r = 5 in $\pi \approx 3.14$
$A_{ring} = \pi\ (10^2 - 5^2) = \pi (100 - 25) = 75\ \pi$
$= 75(3.14) \approx 235.5\ in^2$
Figure 5

Grade 8 Chapter 3 Lesson 8
Circles
Review 1 A
Solve the problems
Use 3.14 for π

1) **Find the diameter of a circle, if its radius equals 15 ft.**

2) **Find the radius of a circle, if its diameter equals 20 in.**

3) **Find the circumference of a circle, if its radius equals 5 yd.**

4) **Find the circumference of a circle, if its diameter equals 10 in.**

5) **Find the diameter of a circle if its circumference equals 314m.**

6) **Find the radius of a circle, if its circumference equals 628 cm.**

7) **Find the circumference of a semicircle, if its diameter is 5 in.**

8) **Find the circumference of a semicircle, if its radius is 10 ft.**

9) **Find the diameter of a semicircle if its circumference is $C = 4\pi$ in**

10) **Find the radius of a semicircle if its circumference is $C = 4\pi$ in**

Grade 8 Chapter 3 Lesson 8

Review 1 B
Solve the problems
Use 3.14 for π

1) Find the area of a circle, if its radius equals 10 yd.

2) Find the area of a circle, if its diameter is 20 m.

3) Find the radius of a circle, if the area equals 314 in^2.

4) Find the diameter of a circle, if the area equals 1256 m^2.

5) Find the area of a circle, if its circumference equals 20π in.

6) Find the circumference of a circle if its area equals 100π ft^2.

7) Find the exact value of the area of a ring, if its big radius is 2.5 cm and its small radius is 1.5 cm.

8) Find the big radius of a ring, if its area is 4π yd, and the small radius is 1.5 yd.

9) Find the small radius of a ring, if its area is 4π cm, and the big radius is 2.5 cm.

10) Find the circumference of a circle, if its diameter is 7 ft and $\pi \approx \frac{22}{7}$

Grade 8 Chapter 3 Lesson 9

Solid Geometry
Unit for volume

1 cm = 10 mm **$1\ cm^3 = 1000\ mm^3$**

1 yd = 3 ft **$1\ yd^3 = 27\ ft^3$** **s is a side**

SA is a surface area **V is a volume** **d is a diagonal**

Cube

The volume of a cube is equal cube of the its side: $V = s^3$

Surface Area of Cube $SA = 6s^2$

Diagonal of Cube is $B_1D = d$ $d = s\sqrt{3}$

Example 1
Find the volume, the surface of area, and the diagonal of cube, if s = 4 ft.

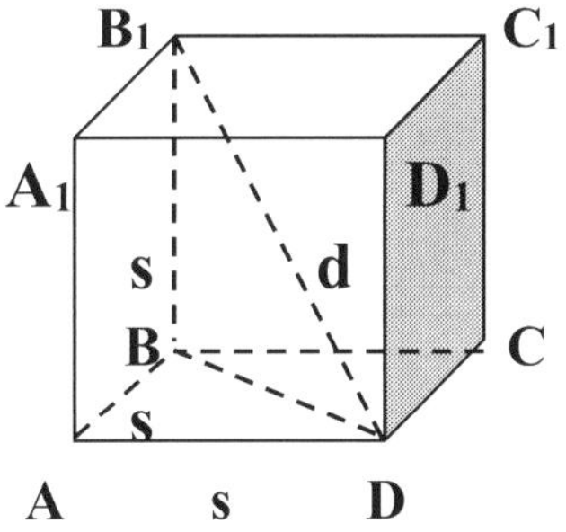

$V = s^3 = 4^3 = 64\ ft^3$

$SA = 6 \cdot 4^2 = 6 \cdot 16 = 96\ ft^2$

$d = s\sqrt{3} = 4\sqrt{3}\ ft$

Figure 1

Rectangular Solid

The volume of a rectangular solid is equal to the product length, width, are height. A rectangular solid has length l, width w, and height h.

Volume of Rectangular Solid $V = lwh$
Surface Area of Rectangular Solid $SA = 2lw + 2lh + 2wh$
Diagonal of Rectangular Solid $d^2 = l^2 + w^2 + h^2$

Grade 8 Chapter 3 Lesson 9
Rectangular solid

Example 2

Find the volume, the surface of area, and the diagonal of rectangular solid, if l = 10 in, w = 6 in, h = 8 in.

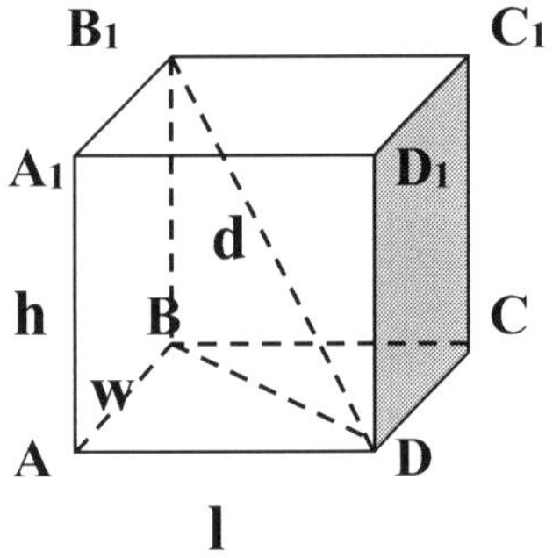

Figure 2

$V = l\,w\,h = 10 \cdot 6 \cdot 8 = 60 \cdot 8 = 480 \text{ in}^3$

$SA = 2lw + 2lh + 2wh = 2 \cdot 10 \cdot 6 + 2 \cdot 10 \cdot 8 + 2 \cdot 6 \cdot 8 =$
$20 \cdot 6 + 20 \cdot 8 + 12 \cdot 8 = 120 + 160 + 96 = 376 \text{ in}^2$

$d^2 = l^2 + w^2 + h^2 = 10^2 + 6^2 + 8^2 = 100 + 36 + 64 = 100 + 100 = 200$
$d = \sqrt{200} = \sqrt{100 \cdot 2} = 10\sqrt{2}$ in

Sphere

The volume of a sphere is equal to the product four over three of the number π by cube of radius. An example of a sphere is any balls.

Volume of sphere is $V = \frac{4}{3}\pi r^3$ **where** $\pi \approx \frac{22}{7}$ **or** $\pi \approx 3.14$

Surface Area of Sphere $SA = 4\pi r^2$ $OB = r$

Example 3

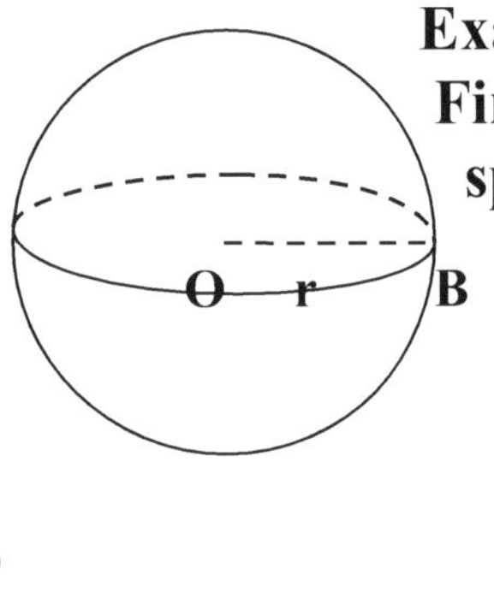

Find exact volume and exact surface area of a sphere if the radius is 3 m.

$V = \frac{4}{3}\pi r^3 = \frac{4}{3}\pi \times 3^3 = \frac{4}{3} \times \frac{27}{1}\pi = 4 \times 9\pi = 36\pi \text{ m}^3$

$SA = 4\pi r^2 = 4\pi \times 3^2 = 4 \times 9\pi = 36\pi \text{ m}^2$

Figure 3

Grade 8 Chapter 3 Lesson 9

Regular Prism

The volume of a regular prism is equal to the product of the base and height:
$V = Base \cdot h$ Base means area of triangles ABC or $A_1B_1C_1$

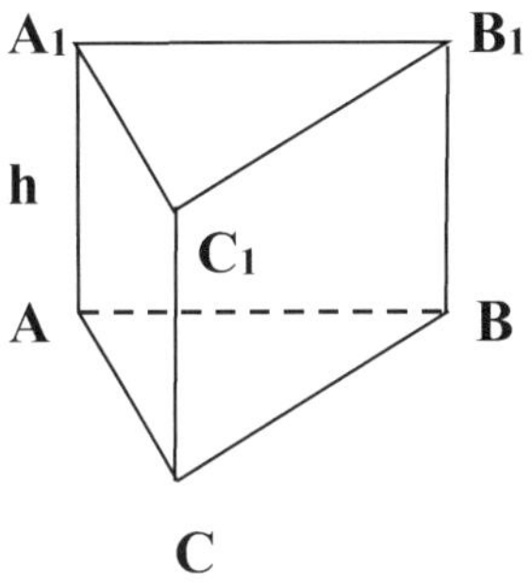

Figure 4

Example 4 Find the volume of a regular prism if the base is 25 m^2 and the height is 10 m. Base = Area of triangle ABC = 25 m^2 h = 10 m
$V = B_{base} \cdot h = 25 \cdot 10 = 250\ m^3$

Right Circular Cylinder

The volume of a right circular cylinder is equal to the product of the number π, and the square of radius by height: $V = \pi r^2 h$

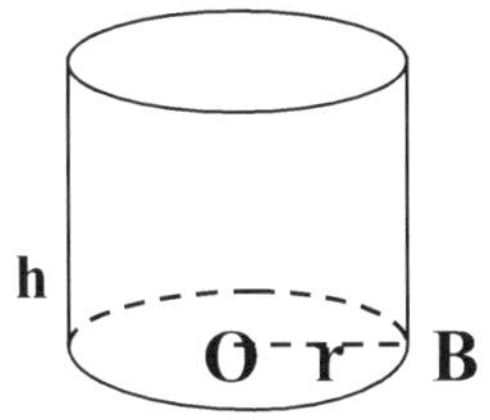

Figure 5
Example 5 Find volume of a right circular cylinder
if r = 7 in h = 10 in and $\pi \approx \frac{22}{7}$

$$V = \pi r^2 h = \frac{22}{7} \cdot 7^2 \cdot 10 = \frac{22}{7} \cdot \frac{49}{1} \cdot 10 = 22 \cdot 7 \cdot 10 = 1540\ in^3$$

Grade 8 Chapter 3 Lesson 9

Regular Pyramid

The volume of a regular pyramid is equal to the product of one-third of base by height: $V = \frac{1}{3}\text{Base} \cdot h$ **Base means area of triangles ABC**

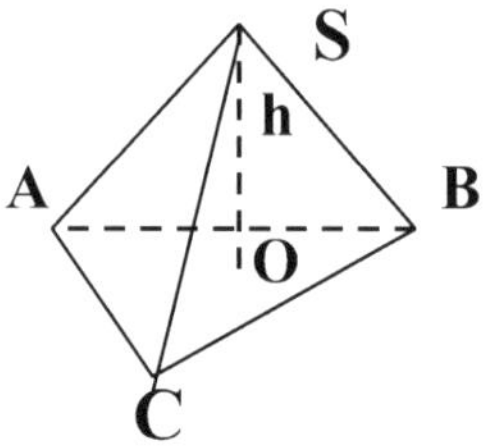

Figure 6

Example 6 **Find the volume of a regular pyramid if the base is 120 m² and height is 5 m. The base is the area of the triangle ABC = 120 m²**

$h = 5$ m $\quad V = \frac{1}{3}\text{Base} \cdot h = \frac{1}{3} \cdot 120 \cdot 5 = 40 \cdot 5 = 200 \text{ m}^3$

Right Circular Cone

The volume of a right circular cone is equal to the product of one-third of square radius by height: $V = \frac{1}{3}\pi r^2 h \quad \pi \approx 3.14 \quad$ or $\pi \approx \frac{22}{7}$

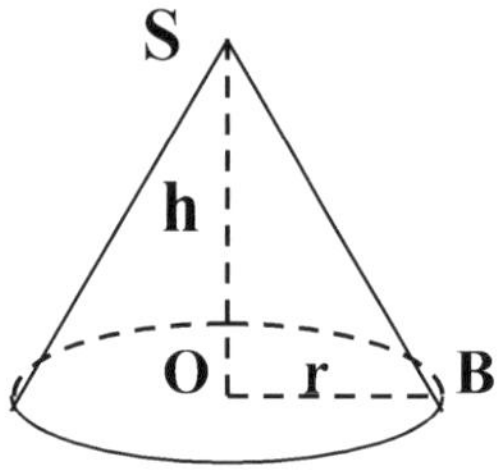

Figure 7

Example 7 **Find the volume of a right circular cone if its height is 10 ft and radius is 3 ft:** $h = 10$ ft $\quad r = 3$ ft

$V = \frac{1}{3}\pi r^2 h = \frac{1}{3}\pi \cdot 3^2 \cdot 10 = \frac{1}{3} \cdot 9 \cdot 10\pi = 3 \cdot 10\pi = 30\pi = 30\,(3.14) = 94.2 \text{ ft}^3$

Grade 8 Chapter 3 Lesson 9

Solid Geometry
Review 1 A
Solve the problems

Cube		$V = s^3$		
1)	$s = 10$ cm	$V = ?$		
2)	$V = 343$ in^3	$s = ?$		
Cube		$SA = 6s^2$		
3)	$s = 10$ in	$SA=?$		
4)	$SA = 600$ cm^2	$s = ?$		
5)	$SA = 600$ ft^2	$V = ?$		
Cube		$d = s\sqrt{3}$		
6)	$s = 10$ ft	$d = ?$		
7)	$d = 20\sqrt{3}$ in	$s = ?$		
Rectangular solid		$V = l\ w\ h$		
8)	$l = 5$ ft	$w = 4$ ft	$h = 2$ ft	$V = ?$
9)	$l = 2.5$ in	$w = 0.8$ in	$V = 8$ in^3	$h = ?$
10)	$l = 1.25$ dm	$h = 0.8$ dm	$V = 4$ dm^3	$w = ?$
11)	$w = 1$ cm	$h = 2$ cm	$V = 10$ cm^3	$l = ?$

Grade 8 Chapter 3 Lesson 9

Review 1 A
Solve the problems

Rectangular solid		$SA = 2lw + 2lh + 2wh$		
12)	$l = 8$ ft	$w = 5$ ft	$h = 4$ ft	$SA = ?$
13)	$l = 8$ in	$w = 3$ in	$SA = 108\ in^3$	$h = ?$
14)	$l = 6$ dm	$h = 4$ dm	$SA = 108\ dm^3$	$w = ?$
15)	$w = 4$ cm	$h = 3$ cm	$SA = 108\ cm^3$	$l = ?$

Rectangular solid		$d^2 = l^2 + w^2 + h^2$		
16)	$l = 6$ ft	$w = 3$ ft	$h = 2$ ft	$d = ?$
17)	$l = 6$ in	$w = 3$ in	$d = 7$ in	$h = ?$
18)	$w = 3$ dm	$h = 2$ dm	$d = 7$ dm	$l = ?$
19)	$l = 6$ cm	$h = 2$ cm	$d = 7$ cm	$w = ?$
20)	$l = 7$ in	$w = 6$ in	$h = 6$ in	$d = ?$

Grade 8 Chapter 3 Lesson 9

Solid Geometry
Review 1 B
Solve the problems

Sphere $V = \frac{4}{3}\pi r^3$ $\pi \approx 3.14$

1) $r = 3$ ft $V = ?$

2) $V = 113.04$ in^3 $r = ?$

3) $d = 6$ cm $V = ?$

4) $V = 113.04$ m^3 $d = ?$

Sphere $SA = 4\pi r^2$

5) $r = 3$ ft $SA = ?$

6) $SA = 1256$ in^2 $r = ?$

Regular Prism $V = \text{Base} \cdot h$

7) Base = 25 m^2 $h = 4$ m $V = ?$

8) Base = 125 in $V = 1000$ in^3 $h = ?$

9) $h = 10$ ft $V = 1000$ ft^3 Base = ?

Grade 8 Chapter 3 Lesson 9

Review 1 B
Solve the problems

Right Circular Cylinder $V = \pi r^2 h$ $\pi \approx 3.14$

10) $r = 10$ in $h = 20$ in $V = ?$

11) $r = 20$ ft $V = 6280\ ft^3$ $h = ?$

12) $h = 10$ cm $V = 3140\ cm^3$ $r = ?$

13) $d = 20$ m $h = 20$ m $V = ?$

14) $h = 20$ yd $V = 6280\ yd^3$ $d = ?$

Regular Pyramid $V = \frac{1}{3} \text{Base} \cdot h$

15) $\text{Base} = 30\ in^2$ $h = 10$ in $V = ?$

16) $\text{Base} = 45\ ft^2$ $V = 300\ ft^3$ $h = ?$

17) $h = 3$ cm $V = 100\ cm^3$ $\text{Base} = ?$

Right Circular Cone $V = \frac{1}{3} \pi r^2 h$ $\pi \approx 3.14$

18) $r = 10$ ft $V = 314\ ft^3$ $h = ?$

19) $h = 3$ cm $V = 314\ cm^3$ $r = ?$

20) $r = 3$ in $h = 6$ in $V = ?$

Grade 8 Chapter 3 Lesson 10

Table 11 Conditional Equations

Six Properties of True Equality: if a = b then

$a + c = b + c$ $\quad a - c = b - c$ $\quad ac = bc$ $\quad a \div c = b \div c$ $\quad c \neq 0$

If $a + c = b + c$ $\quad a = b$ $\quad 3y + 3 = 6 + 3$ $\quad 3y = 6$ $\quad y = 2$

If $a - b = c$ $\quad a = c + b$ $\quad 3y + 2 = 8$ $\quad 3y = 8 - 2$ $\quad 3y = 6$ $\quad y = 2$

If $a + c = b + c$ $\quad a = b$ $\quad 3y - 9 = 0$ $\quad 3y = 0+9$ $\quad 3y = 9$ $\quad y = 3$

If $a + b = c$ $\quad a = c - b$ $\quad 3y + 2 = 8$ $\quad 3y = 8 - 2$ $\quad 3y = 6$ $\quad y = 2$

Conditional equation has one solution $\quad 2y = 8$ $\quad y = 4$

Rule for Solving Equations

1) Open parentheses $a(b + c) = ab + ac$
2) Combine like terms
3) Eliminate number on left side
4) Eliminate letter on right side
5) Multiplication property

Equation with Whole Numbers

$-3(y - 1) = -2(y + 2) + 6$ $\quad -3y + 3 = -2y - 4 + 6$ $\quad -3y + 3 = -2y + 2$

$-3y + 2y = -3 - 2$ $\quad -y = -5$ $\quad y = 5$

Equations with Decimals

$3(y - 5.8) = 6.5(y + 1.2) - 12.25$ $\quad 3y - 17.4 = 6.5y + 7.8 - 12.25$

$3y - 17.4 = 6.5y - 4.45$ $\quad 3y = 6.5y + 12.95$

$-3.5y = 12.95$ $\quad y = -3.7$

Equations with Fraction (× LCD)

$\frac{1}{6} - \frac{y}{4} = \frac{2-y}{3}$ $\quad \frac{12}{1} \times \frac{1}{6} - \frac{12}{1} \times \frac{y}{4} = \frac{12}{1} \times \frac{2-y}{3}$ $\quad 2 - 3y = 4(2 - y)$

$2 - 3y = 8 - 4y$ $\quad -3y + 4y = 8 - 2$ $\quad y = 6$

Proportions

$12 \div 4 = 18 \div 6$ $\quad a \div b = c \div d$ $\quad \frac{12}{4} = \frac{18}{6}$ $\quad 3 = 3$ $\quad \frac{a}{b} = \frac{c}{d}$

$12 \times 6 = 72$ $\quad 4 \times 18 = 72$ $\quad 12 \times 6 = 4 \times 18$ $\quad ad = bc$

$12 \div 18 = 4 \div 6$ $\quad 6 \div 4 = 18 \div 12$ $\quad 6 \div 18 = 4 \div 12$

$\frac{12}{4} = \frac{18}{y}$ $\quad 12y = 4 \cdot 18$ $\quad 12y = 72$ $\quad y = 72 \div 12$ $\quad y = 6$

$\frac{12}{y} = \frac{18}{6}$ $\quad 18y = 12 \cdot 6$ $\quad 18y = 72$ $\quad y = 72 \div 18$ $\quad y = 4$

Grade 8 Chapter 3 Lesson 10

Table 12 Geometry

Right triangles

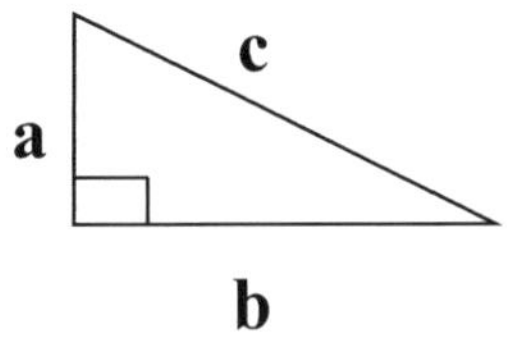

1) Pythagorean Theorem — 1) $a^2 + b^2 = c^2$
2) Area — 2) $A_\Delta = 0.5ab$

Triangles

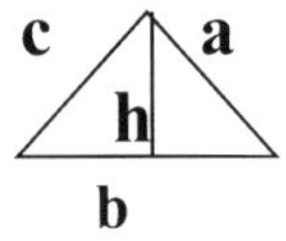

3) Perimeter — 3) $P = a + b + c$
4) Area — 4) $A_\Delta = 0.5bh$

Parallelogram

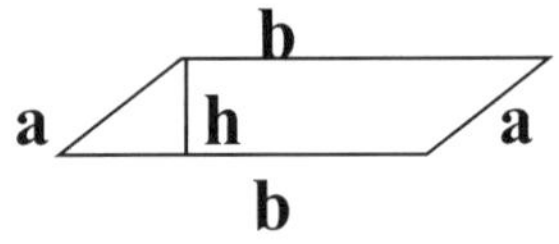

5) Perimeter — 5) $P_{par} = 2a + 2b$
6) Area — 6) $A_{par} = b\ h$

Rhombus

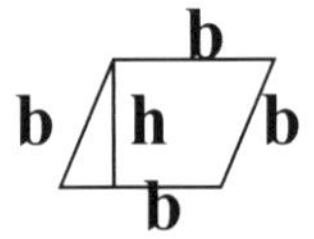

7) Perimeter — 7) $P_{rh} = 4b$
8) Area — 8) $A_{rh} = b \cdot h$

Rectangle

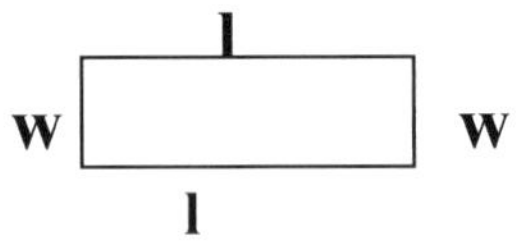

9) Perimeter — 9) $P_{rec} = 2l + 2w$
10) Area — 10) $A_{rec} = l \cdot w$

Square

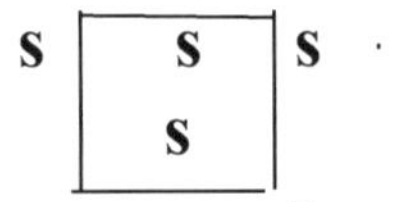

11) Perimeter — 11) $P_{\blacksquare} = 4s$
12) Area — 12) $A_{\blacksquare} = s^2$

Trapezoid

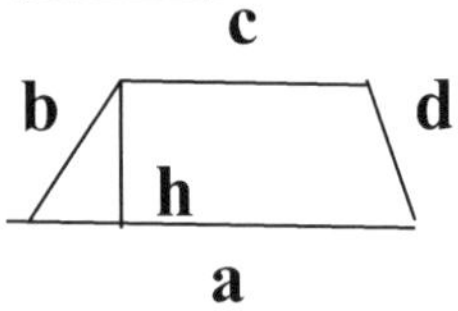

13) Perimeter — 13) $P_{tr} = a + b + c + d$
14) Area — 14) $A_{tr} = 0.5\ h\ (a + c)$

Circle

15) Number π in fraction $\approx \frac{22}{7}$ 16) Number π in decimal ≈ 3.14

17) Diameter $d = 2r$ 18) Circumference 1 (d) $C = \pi d$

19) Circumference 2 (r) $C = 2\pi r$ 20) Area $A_c = \pi r^2$

Solid Geometry

Cube 1) $V = s^3$ 2) $SA = 6s^2$ 3) $d = s\sqrt{3}$ Sphere 4) $V = \frac{4}{3}\pi r^3$ 5) $SA = 4\pi r^2$

Rectangular Solid: 6) $V = l\ w\ h$ 7) $SA = 2lw + 2lh + 2wh$ 8) $d^2 = l^2 + w^2 + h^2$

Regular Prism 9) $V = \text{Base} \cdot h$ Regular Pyramid 10) $V = \frac{1}{3}\text{Base} \cdot h$

Right Cylinder 11) $V = \pi r^2 h$ Right Cone 12) $V = \frac{1}{3}\pi r^2 h$

Memorize Formulas of Geometry

Grade 8 Chapter 3 Lesson 10

Mental Math 15
Law for Fractions

1) $\frac{1}{2}\times8+\frac{1}{2}\times2$

2) $\frac{4}{3}\times6-\frac{1}{3}\times6$

3) $\frac{1}{4}\times8+\frac{3}{4}\times8$

4) $\frac{6}{5}\times5-\frac{1}{5}\times5$

5) $\frac{1}{3}\times9+\frac{2}{3}\times9$

6) $\frac{3}{2}\times8-\frac{1}{2}\times8$

7) $\frac{1}{3}\times6+\frac{2}{3}\times6$

8) $\frac{5}{4}\times4-\frac{1}{4}\times4$

9) $\frac{1}{5}\times5+\frac{4}{5}\times5$

10) $\frac{7}{6}\times6-\frac{1}{6}\times6$

11) $\frac{1}{6}\times6+\frac{5}{6}\times6$

12) $\frac{5}{2}\times4-\frac{1}{2}\times4$

13) $\frac{3}{7}\times7+\frac{4}{7}\times7$

14) $\frac{7}{3}\times3-\frac{1}{3}\times3$

15) $\frac{3}{8}\times8+\frac{5}{8}\times8$

16) $\frac{5}{4}\times8-\frac{1}{4}\times8$

17) $\frac{4}{9}\times9+\frac{5}{9}\times9$

18) $\frac{7}{5}\times5-\frac{2}{5}\times5$

19) $\frac{3}{5}\times5+\frac{2}{5}\times5$

20) $\frac{7}{6}\times6-\frac{1}{6}\times6$

Grade 8 Chapter 3 Lesson 10

Mental Math 16

Exponents

1) 0^2	11) 10^2
2) 1^2	12) 20^2
3) 2^2	13) 30^2
4) 3^2	14) 40^2
5) 4^2	15) 50^2
6) 5^2	16) 60^2
7) 6^2	17) 70^2
8) 7^2	18) 80^2
9) 8^2	19) 90^2
10) 9^2	20) 100^2

Grade 8 Chapter 3 Lesson 11
Reviews for Chapter 3
Review 3 A

Date	Name
Read directions carefully.	Answers must be in the indicated place.
Show all steps for full credits.	Do only 1 operation per 1 step.
Each problem is worth 5 points.	Solve 1 problem 5 minutes.

1) Find one leg of a right triangle if its hypotenuse is 9.5 ft and another leg is 5.7 ft. Draw the picture.

1) ______

2) Find the area of a right triangle, if its legs are $5\frac{5}{6}$ m and $4\frac{4}{5}$ m.

Draw the picture.

2) ______

3) Find the area of a triangle if it sides are 5 cm, 12 cm, and 13 cm. Draw the picture.

3) ______

4) Find the area of a square if it side is $1\frac{1}{2}$ ft. Draw the picture.

4) ______

5) Find the area of a rhombus if the base is $3\frac{1}{3}$ in and the height is $2\frac{1}{4}$ in. Draw the picture.

5) ______

Grade 8 Chapter 3 Lesson 11
Review 3 A

Date | **Name**

Read directions carefully. | **Answers must be in the indicated place.**

Show all steps for full credits. | **Do only 1 operation per 1 step.**

Each problem is worth 5 points. | **Solve 1 problem 5 minutes.**

Solve the problems

6) Find the area of a trapezoid; if the height is $6\frac{2}{3}$ cm, and bases are $9\frac{1}{2}$ cm and $8\frac{1}{2}$ cm. Draw the picture.

6) ______

7) Find the circumference of circle that has a diameter is 28 in and $\pi \approx \frac{22}{7}$. Draw the picture.

7) ______

8) Find the area of a circle if its radius r = 10 m and $\pi \approx 3.14$ Draw the picture.

8) ______

9) Find the volume of a cube if its side is $1\frac{1}{2}$ in.

9) ______

10) Find the exact value volume of a sphere if radius $r = 1\frac{1}{2}$ ft.

10) ______

Grade 8 Chapter 3 Lesson 11
Review 3 B

Date
Name

Read directions carefully.
Answers must be in the indicated place.
Show all steps for full credits.
Do only 1 operation per 1 step.
Each problem is worth 5 points.
Solve 1 problem 5 minutes.

1) Find one leg of a right triangle if its hypotenuse is 6.5 ft and the other leg is 5.2 ft. Draw the picture.

1) ______

2) Find the area of a right triangle, if it legs are $6\frac{1}{4}$ m and $4\frac{4}{5}$ m. Draw the picture.

2) ______

3) Find the area of a triangle if its sides are 9 cm, 12 cm, and 15 cm Draw the picture.

3) ______

4) Find the area of a rectangle if its length is $3\frac{1}{3}$ ft and its width is $2\frac{2}{5}$ ft. Draw the picture.

4) ______

5) Find the area of a parallelogram if the base is $5\frac{3}{5}$ in and the height is $3\frac{3}{4}$ in. Draw the picture.

5) ______

Grade 8 Chapter 3 Lesson 11
Review 3 B

Date
Name
Read directions carefully.
Answers must be in the indicated place.
Show all steps for full credits.
Do only 1 operation per 1 step.
Each problem is worth 5 points.
Solve 1 problem 5 minutes.

Solve the problems

6) Find the area of a trapezoid, if its height is $5\frac{1}{3}$ cm, and bases are $6\frac{1}{2}$ cm and $5\frac{1}{2}$ cm. Draw the picture.

6) ______

7) Find the area of a circle if its radius 7 in and $\pi = \frac{22}{7}$. Draw the picture.

7) ______

8) Find the area of a ring of a circle is its radii are 2.5 in and 1.5 in and $\pi \approx 3.14$. Draw the picture.

8) ______

9) Find the volume of a rectangular solid if its length is 4 in, width is 2.5 in and height is 0.5 in

9) ______

10) Find the exact value volume of a sphere if radius r = $4\frac{1}{2}$ ft.

10) ______

Grade 8 Chapter 3 Lesson 12
Tests for Chapter 3
Test 3 A

Date | **Name**
Read directions carefully. | **Answers must be in the indicated place.**
Show all steps for full credits. | **Do only 1 operation per 1 step.**
Each problem is worth 5 points. | **Solve 1 problem 5 minutes.**

1) Find one leg of a right triangle if its hypotenuse is $7\frac{1}{2}$ ft and the other leg is $7\frac{1}{5}$ ft. Draw the picture.

1) ______

2) Find the area of a right triangle, if its legs are $6\frac{2}{3}$ m and $4\frac{1}{2}$ m. Draw the picture.

2) ______

3) Find the area of a triangle if its sides are 12 cm, 16 cm, and 20 cm. Draw the picture.

3) ______

4) Find the area of a square if its side is $2\frac{1}{2}$ ft. Draw the picture.

4) ______

5) Find the area of a rhombus if the base is $7\frac{1}{2}$ in and the height is $4\frac{2}{3}$ in. Draw the picture.

5) ______

Grade 8 Chapter 3 Lesson 12
Test 3 A

Date | **Name**
Read directions carefully. | **Answers must be in the indicated place.**
Show all steps for full credits. | **Do only 1 operation per 1 step.**
Each problem is worth 5 points. | **Solve 1 problem 5 minutes.**

Solve the problems

6) **Find the area of a trapezoid, if the height is $2\frac{2}{3}$ cm, and the bases are $6\frac{1}{2}$ cm and $5\frac{1}{2}$ cm. Draw the picture.**

6) ______

7) **Find the circumference of a circle that has a diameter of 14 in and $\pi = \frac{22}{7}$. Draw the picture.**

7) ______

8) **Find the area of a circle if radius r = 20 m and $\pi \approx 3.14$ Draw the picture.**

8) ______

9) **Find the volume of a cube if its side is $2\frac{1}{2}$ in.**

9) ______

10) **Find the exact value volume of a sphere if radius r = 3 ft.**

10) ______

Grade 8 Chapter 3 Lesson 12
Test 3 B

Date | **Name**

Read directions carefully.
Show all steps for full credits.
Each problem is worth 5 points.
Answers must be in the indicated place.
Do only 1 operation per 1 step.
Solve 1 problem 5 minutes.

1) Find one leg of a right triangle if its hypotenuse is $5\frac{1}{10}$ ft and the other leg is $2\frac{2}{5}$ ft. Draw the picture.

1) ______

2) Find the area of a right triangle, if its legs are $9\frac{3}{4}$ m and $5\frac{1}{3}$ m . Draw the picture.

2) ______

3) Find area of a triangle if its sides are 15 cm, 20 cm, and 25 cm Draw the picture.

3) ______

4) Find the area of a rectangle if its length is $10\frac{1}{2}$ ft and its width is $5\frac{1}{3}$ ft. Draw the picture.

4) ______

5) Find the area of a parallelogram if the base is $7\frac{1}{2}$ in and the height is $6\frac{2}{3}$ in. Draw the picture.

5) ______

Grade 8 Chapter 3 Lesson 12
Test 3 B

Date | **Name**

Read directions carefully. **Answers must be in the indicated place.**
Show all steps for full credits. **Do only 1 operation per 1 step.**
Each problem is worth 5 points. **Solve 1 problem 5 minutes.**

Solve the problems

6) Find the area of a trapezoid, if its height is $3\frac{1}{3}$ cm, and the bases are $9\frac{3}{4}$ cm and $8\frac{1}{4}$ cm. Draw the picture.

6) ______

7) Find the area of a circle if the radius is 14 in and $\pi \approx \frac{22}{7}$.

Draw the picture.

7) ______

8) Find the area of a ring of a circle if its radii are 6.5 in and 5.5 in and $\pi \approx 3.14$. Draw the picture.

8) ______

9) Find the volume of a rectangular solid if its length is 12.5 in, its width is 5 in and its height is 4 in

9) ______

10) Find the exact value volume of a sphere if the radius r = 6 ft.

10) ______

Grade 8 Answers Chapter 3

Lesson 1 1) 30 cm^2 2) 616 in^2 3) 37.68 m^2 4) 250 in^3 5) 288π ft^3

Lesson 2 Segments. Review 1 A

1) a) > b) < c) = 2) a) 10cm b) 2 cm 3) 12 cm 4) 8 in 5) 2 in
6) 5 in 7) 6 cm 8) 8 in 9) 4 cm 10) 7 in

Review 1B

1) 10 in 2) 15 cm 3) 6 in 4) 6 cm 5) 7 in
6) 5 in 7) 6 cm 8) 4 cm 9) 8 in 10) 3 in

Lesson 3 Angles. Review 1 A

1) 75^0 2) 135^0 3) 90^0 4) 150^0 5) 45^0
6) 60^0 7) 30^0 8) 120^0 9) 30^0 10) 45^0

Review 1 B

1) 150^0, 30^0 2) 100^0, 80^0 3) 81^0, 99^0 4) 84^0, 96^0 5) 18^0, 162^0
6) 30^0, 150^0 7) 80^0, 100^0 8) 70^0, 110^0 9) 36^0, 144^0 10) 80^0, 100^0

Lesson 4 Triangles. Review 1 A

1) 90^0 2) 65^0 3) 45^0 4) 60^0 5) 18 cm
6) 10 m 7) 36 in 8) 5 ft 9) 16 yd 10) 18 ft

Review 1 B 1) 8 in 2) 50 cm 3) 0.6 in 4) 75 cm
5) 0.8 ft 6) 8, 16, 16 cm 7) 12, 12, 2 in
8) 6, 6, 8 in 9) 10, 15, 15 in 10) 8, 8, 6 cm

Lesson 5 Right Triangles. Review 1 A

1)	3	2)	8	3)	15	4)	12	5)	20
6)	30	7)	21	8)	32	9)	45	10)	30
11)	44	12)	60	13)	39	14)	56	15)	75
16)	48	17)	68	18)	90	19)	57	20)	80
21)	13	22)	10	23)	36	24)	52	25)	25

Review 1 B

1)	7.2	2)	2.5	3)	1.4	4)	7.2	5)	10
6)	2	7)	4.2	8)	1.7	9)	1.6	10)	4.5
11)	4.1	12)	1.8	13)	6	14)	8.5	15)	6
16)	7.2	17)	5	18)	2.1	19)	9.6	20)	5.8
21)	0.8	22)	3	23)	5.1	24)	1.8	25)	8.4

Lesson 6 Perimeter. Review 1 A

1)	24 cm	2)	15 in	3)	10 ft	4)	28 cm
5)	20 yd	6)	10 in	7)	40 dm	8)	6 cm
9)	20 cm	10)	12 in	11)	14 ft	12)	36 m
13)	20 ft	14)	28 cm	15)	50 ft	16)	10 yd
17)	24 cm	18)	30 ft	19)	10 m	20)	32 in

Grade 8 Answers Chapter 3

Lesson 7 Area. Review 1 A

1)	64 in^2	2)	5 cm	3)	36 yd^2	4)	9 cm
5)	11 ft	6)	12 m	7)	10 in	8)	10 m
9)	12 in	10)	8 ft	11)	40 m^2	12)	20 m
13)	40 in^2	14)	10 m	15)	6 m^2	16)	24in^2
17)	30ft^2	18)	54 cm^2	19)	84yd^2	20)	96dm^2

Lesson 8 Circle. Review 1 A

1) 30 ft 2) 10 in 3) 31.4 yd 4) 31.4 m 5) 100m
6) 100 cm 7) 7.85 in 8) 314 ft 9) 8 in 10) 4 in

Review 1 B

1) 314 yd^2 2) 314 m^2 3) 10 in 4) 40m 5) 314in^2
6) 62.8 ft 7) 4π cm^2 8) 2.5 yd 9) 1.5 cm 10) 22ft

Lesson 9 Solid Geometry. Review 1 A

1) 1000 dm^3 2) 7 in 3) 600 in^2 4) 10cm 5) 1000 ft^3
6) $10\sqrt{3}$ ft 7) 20 in 8) 40 ft^3 9) 4 in 10) 4 dm
11) 5 cm 12) 184 ft^2 13) 4 in 14) 3dm 15) 6 cm
16) 7 ft 17) 2 in 18) 6dm 19) 3cm 20) 11in

Review 1 B

1) 113.04 ft^3 2) 3in 3) 113.04 cm^3 4) 6 m 5) 113.04 ft^2
6) 10in 7) 100 m^3 8) 8 in 9) 100 ft^2 10) 6280 in^3
11) 5 ft 12) 10 cm 13) 6280 m^3 14) 20 yd 15) 100 in^3
16) 20 ft 17) 100 cm^2 18) 3 ft 19) 10 cm 20) 56.52 in^3

Lesson 11 Review 3 A

1) 7.6 ft 2) 14 m^2 3) 30 cm^2 4) 2.25 ft^2 5) 7.5 in^2
6) 60 cm^2 7) 88 in 8) 314 m^2 9) 3.375 in^3 10) 4.5π ft^3

Review 3 B

1) 3.9 ft 2) 15 m^2 3) 54 cm^2 4) 8 ft^2 5) 21 in^2
6) 32 cm^2 7) 154 in^2 8) 12.56 m^2 9) 5 in^3 10) 121.5π ft^3

Lesson 12 Test 3 A

1) 2.1 ft 2) 15 m^2 3) 96 cm^2 4) 6.25 ft^2 5) 35 in^2
6) 16 cm^2 7) 44 in 8) 1256 m^2 9) 15.625 in^3 10) 36π ft^3

Test 3 B

1) 4.5 ft 2) 26 m^2 3) 150 cm^2 4) 56 ft^2 5) 50 in^2
6) 30 cm^2 7) 616 in^2 8) 37.68 m^2 9) 250 in^3 10) 288π ft^3

Grade 8

Chapter 4

Word

Problems

Grade 8 Chapter 4 Contents
Word Problems
Lesson 1 – 12
Basic Problems

Work with lesson 1, read rules and examples, 2) Solve the problems from contents, 3) Check answers in the end of chapter 1, 4) Work with the next lessons 2, 3, 4, 5, …, 5) Solve Test 1 A, 6) Grading Scale: A =10 points, B=9points, C=8 points, D=7 points, F=0-6 points, 7) If you solve 10 problems correctly, you get an A, 8) If have an A, you can move on to next chapter, 9) If your grade is lower than an A, work with the mistakes, then solve Test 1B, 10) When you finish all of the chapters, solve Final Exam in last chapter.

Grade 8 Chapter 4 Lesson 1
Table Word Problems

1) Decimals

a) To find the amount by known base, multiply the base by a decimal:

Find the amount 0.6 of 40 $40 \times 0.6 = 24$

b) To find the base by known amount, divide the amount by a decimal: Finding the base: 0.6 is equivalent \$ 24 $24 \div 0.6 = 40$

c) To find the ratio between two numbers, divide the numbers.

Find the ratio 24 from 40 $24 \div 40 = 0.6$

2) Fractions

a) To find the amount by known base, multiply the base by a fraction.

Finding the amount: $\frac{3}{5}$ of 40 $40 \times \frac{3}{5} = \$24$

b) To find the base by known amount, divide amount by a fraction.

Finding the base: $\frac{3}{5}$ equivalent 24 $24 \div \frac{3}{5} = \$40$

c) To find the ratio between two numbers, divide the numbers.

Find the ratio: 24 from \$40 $24 \div 40 = \frac{24}{40} = \frac{3}{5}$

3) Uniform Motion: Distance - d Time - t Rate – r (Speed)

a) To find distance, multiply rate by time: d = rt

What distance did a man go for 2 hours if his rate was 5 mi/hr?

$d = 5 \cdot 2 = 10$ mi

b) To find time, divide distance by speed. $t = d \div r$

How much time does a man needs if he runs 8 mi with a rate 4 mi/hr?

$t = 8 \div 4 = 2$ hr

c) To find speed, divide distance by time. $r = d \div t$

What is the rate a girl has, if she runs 6 mi for 2 hours

$r = 6 \div 2 = 3$ mi / hr

4) Directly proportional quantities

If 4 apples sell for \$1.32. How much do 11 apples cost?

$132 \div 4 = 33$ c $33 \times 11 = 363$ c = \$3.63

Grade 8 Chapter 4 Lesson 1
Pretest

1) A bus drove 46 mi at a speed of 20 mi/h.
   How much time it spent?

1) ______

2) What speed does a boy have, if he drove 65 mi in 4 hours?

2) ______

3) Of 24 children, 62.5% were boys. How many were girls?

3) ______

4) A factory has 6 new machines and 60% were old machine.
   How many machines are in the factory?

4) ______

5) In 75 minutes of lesson, students solve an equation in 30 minutes.
   What percent of lesson time did students use to solve the equation?

5) _____

Grade 8 Chapter 4 Lesson 2

Word Problems on Operations

a) Straight problem
First girl has 3 apples
Second girl has 7apples
Sum of apples is x apples
x = 3 + 7 = 10 apples

b) Opposite problem
1 girl has 3 apples
2 girl has y apples
Sum of apples is 10 apples
y = 10 – 3 = 7 apples

c) Straight problem
I have $10
I spend $ 4
Left $ x
x = 10 – 4 = $ 6

d) Opposite problem
I have $ y
I spend $ 4
Left $ 6
y = 4 + 6 = $ 10

Review 1 A

Solve the Word Problems

1) There are 35 flowers. 24 more are planted. How many are there total?

2) There are 30 red roses and 59 white roses. How many flowers are there in total?

3) Amy bought a candle for 25 cents and napkins for 44 cents. How much did she spend?

4) There are 28 birds. 30 more come. How many are there now?

5) Olivia bought a pencil for 12 cents, a pen for 34 cents, and notebooks for 42 cents. How much did she spend?

6) Ann has 16 pencils. Bob has 14 pencils. How many more does Ann have than Bob?

7) There are 18 balloons. Two of the balloons broke. How many are left?

8) Bob have 64 roses. 53 are red. The rest are white. How many white ones are there?

9) A store has 88 hammers. The store sold 64. How many hammers does the store have now?

10) Lisa has 74 cents. She spent 34 cents. How much does she have left?

Grade 8 Chapter 4 Lesson 2
Review 1 B
Solve the Word Problems on Operations

1) Father works 40 hours each week.
   How many hours does he work in 2 weeks?
2) There are 8 rows of seats in the balcony.
   There are 35 seats in each row.
   How many seats are in the balcony?

3) A student-spelling book has 25 new words on each page.
   There are 9 pages in the first section.
   How many new words are there in the first section?
4) There are 4 classrooms on the first floor.
   Each classroom has 30 seats.
   How many seats are on the first floor?

5) Mother works 8 hours each day.
   How many hours will she work in 5 days?
6) Michele has 63 peaches. She puts 7 peaches in a package.
   How many packages will she have?

7) David worked the same number of hours each day.
   He worked 21 hours in 7 days.
   How many hours did he work each day?
8) One side of a building has 24 windows.
   Each floor has 6 windows on that side.
   How many floors does the building have?

9) Mary has 28 balloons for a party.
   She will give each person 4 balloons.
   How many people will receive balloons?
10) The 45 students in a class separated into 5 groups.
   Each group has the same number of students.
   How many are students in each group?

Grade 8 Chapter 4 Lesson 2
Review 1 C
Solve the Word Problems

1) Al had 32 markers. He bought 14 more. How many are there now?

2) Rachel had 26 nickels. She got 43 more. How many does she have now?

3)There are 9 black pens and 4 blue pens. How many more black pens there than blue pens?

4) 24 cups are on the table but 2 are broken. How many are not broken?

5) The teacher gave each student 2 sheets of paper. There were 25 students. How many sheets of paper did he use?

6) There are 25 baseball players on each team. How many players are there total on 2 teams?

7) Michael bought 2 packages of paper. Each package had 30 sheets. How many sheets of paper did he buy?

8) A farmer planted 54 trees in 9 rows. Each row had the same numbers of trees. How many trees were in each row?

9) Yuri worked 20 hours in 5 days. He worked the same numbers of hours each day. How many hours did he work each day?

10) Olivia read 35 pages of a book in 7 days. She read the same number of pages each day. How many pages did she read each day?

Grade 8 Chapter 4 Lesson 3

Word Problems on Worth

Measurement of worth - \$ 1, 1c

To solve word problems at worth, reduce problem to 1 unit.

How many does cost 15m silk if 8 m of silk cost \$ 32?

$\$32 \div 8 = \$4$ $\$4 \times 15 = \$60$

Review 1 A

Solve the Word Problems

1) Amanda bought 5 bags for 2 c for each. How much money did she pay?

2) Monica has 4 coins of 5 c each. How much money does she have?

3) 1 lb of cabbage cost 10 c. How much money does 3 lbs. of cabbage cost?

4) Nicole bought a journal for 30 c and a book, which cost 3 times more than the journal. How much did the book cost?

5) Sophia bought 24 postcards for 3 c each. How much did she spend?

6) A cap cost \$6 and a coat cost 10 times more than the cap. How much did the coast coat?

7) Amy bought 3 meters of fabric for \$4 for a meter. She has \$38 remaining. How much money did she have?

8) In the park are planted 4 rows of roses with 5 plants in each row. How many roses planted?

9) Mother spilled 2 liters of milk in each of 4 cans. How many liters of milk does mother have in all?

10) Three boxes of oranges, 4 lb. each, brought into the restaurant. How many pounds of oranges brought into the restaurant?

Grade 8 Chapter 4 Lesson 3
Review 1 B
Solve the Word Problems on Worth

1) **Rachel needs to sew 4 buttons onto one coat.**
   How many buttons does the woman need to sew onto 2 coats?

2) **A house has 5 floors and each floor has a height of 4 meters.**
   How many meters high is this house?

3) **A chair cost $40 and table cost 6 times more than the chairs.**
   How much did the table cost?

4) **One fork costs 30 c. How much a set of 3 would forks cost?**

5) **One balls costs $2. How much do 6 balls cost?**

6) **Jenna bought 2 ft of silk for $8. How much money does cost 1 ft of silk cost?**

7) **Linda paid 14 c for 2 packets of salt.**
   How much money does 1 packet of salt cost?

8) **April bought on $1.20 envelopes by 6 c for each.**
   How many envelopes could she buy?

9) **Tiffany bought a dress for $27 and shoes, which cost 3 times less than the dress. How much did the shoes cost?**

10) **Alana paid for $24 for 1 meter of wool and $6 for 1 meter of silk.**
    How many times costlier is the wool than the silk?

Grade 8 Chapter 4 Lesson 3
Review 1 C
Solve the Word Problems

1) A book cost $20. How many books can Jackie buy for $140?

2) Brett has 60c. How many tickets to the movie can he buy for this money, if 1 ticket costs 10 c?

3) Mark divided 6 balls by 2 balls for each boy. How many boys were there?

4) Amy laid out 18 pencils from 3 boxes equally. How many pencils were in each box?

5) Claire divided 12 tulips into bouquets with 4 tulips in each. How many bouquets are there?

6) Carl loaded 16 logs on 2 trucks equally. How many logs did they load in each truck?

7) Andrew tied 10 carrots in bunches with 5 carrots in each. How many bunches are there?

8) On the first day, 8 bags were sold for $32. On the second day 6 of the same bags were sold. How much money paid on the second day?

9) A set of towels cost $1 and 3 packs of napkins cost 75 c. How many packs of napkin can bought instead of a set of towels?

10) A table cost $240 and a chair cost 6 times less than the table. How much did the table and chair cost together?

Grade 8 Chapter 4 Lesson 4

Word Problems on Decimals

In word problems in decimal each word has special meanings
"Base" (B) represents "Whole Quantity",
"Amount" (A) usually represents "Portion of Whole Quantity"

Finding the Amount by Known Base

Amy had whole money \$40. She spent 0.6 of all her whole money.
How much amount of money did she spend? To find how much money did she spent, multiply the base, \$40, by the decimal 0.6.

0.6 of \$40 $\$40 \times 0.6 = \$24$

To find the amount of money by known base, multiply the base by a decimal.

Finding the Base by Known Amount

Amy spent 0.6 of all her whole money; this is equals \$24. How much of whole money did she have before? To find how much whole money she had, divide amount of the money, \$24, by the decimal 0.6.

0.6 equivalent \$24 $\$24 \div 0.6 = 240 \div 6 = \$40$

To find the base by known amount, divide the amount by the decimal.

Find the Ratio

Amy has \$40. She spent \$24. What part of the money did she spent?
To find the ratio between the money she spent to the whole of her money, divide the money she spent, \$24, by her whole money, \$40.
Write ratio as a decimal.

\$24 from \$40 $\$24 \div \$40 = 0.6$

To find ratio between two numbers, divide the numbers.

Average

The quotient of division of the sum of numbers by the number of addends called an average.
Example: Find the average of numbers 7.8, 8.4, 9.3
$(7.8 + 8.4 + 9.3) \div 3 = 25.5 \div 3 = 8.5$

Grade 8 Chapter 4 Lesson 4
Review 1 A
Solve the Word Problems for Decimals

1) In class there were 35 students. Students who can swim are 0.2 of the whole class. How many students can swim?

2) A book has 60 pages. A student reads 0.5 of the whole book. How many pages does he read?

3) Andrew has 15 stamps. He gave his brother 0.2 of all of the stamps. How many stamps did the boy gave to his brother?

4) A tailor prepared 8 dresses. He sewed 0.5 of all of the dresses. How many dresses did he sew?

5) A kindergarten has 36 children. Girls made up 0.25 of all of the children. How many girls were in the kindergarten?

6) A notebook has 24 pages. A girl wrote in 0.625 of the whole notebook. How many pages remain?

7) A book cost $18, and it equals 0.3 of my whole money. How much money do I have?

8) Anna ate 6 plums; that equals 0.6 of all of the plums. How many plums did I have?

9) Briana read 10 pages; that equals 0.4 of all of the pages in the book. How many pages are in the book?

10) In the class, 8 students can dance; this is 0.25 of the whole class. How many students are in class?

Grade 8 Chapter 4 Lesson 4
Review 1 B
Solve the Word Problems

1) Lucy ate 5 of 10 apples. What part of apples did she eat? Write ratio as a decimal.

2) April read 20 pages. What part of the book did she read, if the book has 50 pages? Write ratio as a decimal.

3) Ben has 24 pencils. He gave his friend 6 pencils. What part of pencils did he gave to his friend? Write ratio as a decimal.

4) Lisa has \$36. She spent \$9. What part of the money did she spends? Write ratio as a decimal.

5) The shop had 900 lbs. of sugar. On the first day, they sold 0.3 of this sugar, and on the second day, they sold the same. How many pounds did they sell in the two days together?

6) A book has 80 pages. David read 0.25 of the whole book. How many pages more does he have left to read?

7) A class had 40 students. The girls made up 0.625 of the whole class. How many boys were in the class?

8) Mother has \$10. She used 0.6 of her whole money to buy 4 loaves of bread. What much is one bread worth?

9) Find the average for the numbers: 1.5, 2.2, 3.2?

10) Find the average for the numbers: 4.5, 5.2, 6.2, 8.1?

Grade 8 Chapter 4 Lesson 5

Word Problems on Fractions

In word problems in fraction each word has special meanings
"Base" (B) represents "Whole Quantity",
"Amount" (A) usually represents "Portion of Whole Quantity"

Finding Amount by Known Base

Amy had whole money \$40. She spent $\frac{3}{5}$ of her whole money. How much amount money did she spend? To find how much money did she spent, multiply the base, \$40, by the fraction $\frac{3}{5}$:

$\frac{3}{5}$ of \$40 $\quad\quad$ **$\$40 \times \frac{3}{5} = \$24$**

To find amount by the known base, multiply the base by a fraction or divide the base by denominator; after this, multiply the result at the numerator:
$\$40 \div 5 \times 3 = \$8 \times 3 = \$\ 24$

Finding Base by Known Amount

Amy spent $\frac{3}{5}$ of her whole money; this is equals \$24. How much whole money did she have before? To find how much whole money she had, divide amount of the money, \$24, by the fraction $\frac{3}{5}$:

$\frac{3}{5}$ equivalent \$24 $\quad\quad$ **$\$24 \div \frac{3}{5} = 24 \times \frac{5}{3} = 24 \div 3 \times 5 = \$40$**

To find the base by known amount, divide amount by the fraction or multiply amount by the reciprocals for the fraction.

Find the Ratio

Amy has \$40. She spent \$24. What part of the whole money did she spent?
To find the ratio between the money she spent to the whole of her money, divide the money she spent, \$24, by her whole money, \$40.
Write ratio as a reduced fraction.

\$24 from \$40 $\quad\quad$ **$\$24 \div \$40 = \frac{24}{40} = \frac{3}{5}$**

To find ratio between two numbers, divide the numbers.

Grade 8 Chapter 4 Lesson 5
Review 1 A
Solve the Word Problems for Fractions

1) A class has 35 students. Students, who can swim make up $\frac{1}{5}$ of the whole class. How many students can swim?

2) A book has 60 pages. Student read $\frac{1}{2}$ of the whole book. How many pages did he read?

3) Andrew has 15 stamps. He gave his brother $\frac{1}{5}$ of all of the stamps. How many stamps did Andrew give his brother?

4) A tailor prepared 8 dresses. He sewed $\frac{1}{2}$ of all of the dresses. How many dresses did he sew?

5) A kindergarten has 36 children. The girls make up $\frac{1}{4}$ of the whole children. How many girls are in the kindergarten?

6) A notebook has 24 pages. Amanda wrote in $\frac{5}{8}$ of the whole notebook. How many pages remain?

7) A shop has 900 lb of sugar. On the first day, they sold $\frac{3}{10}$ of all sugar. On the second day, they sold the same amount. How many pounds did they sell in the two days together?

8) A book has 80 pages. Mike read $\frac{1}{4}$ of the whole book. How many pages more does he have left to read?

9) A class has 40 students. The boys' make-up $\frac{5}{8}$ of the whole class. How many girl are in the class?

10) Mother has $10. She used $\frac{3}{5}$ of her whole money to buy 4 loaves bread. How much is one bread worth?

Grade 8 Chapter 4 Lesson 5
Review 1 B
Solve the Word Problems for Fractions

1) A book cost \$18; this it equals $\frac{3}{10}$ of the whole money. How much is the whole money?

2) Sara ate 6 plums; this is equals $\frac{3}{5}$ of all of the plums. How many plums did she have?

3) Brandon read 10 pages; this it equals $\frac{1}{10}$ of the whole pages of the book. How many pages were in the book?

4) In the class, 10 students can dance; this is equals $\frac{1}{4}$ of the whole class. How many students are in the class?

5) Casey ate 5 of 10 apples. What part of apples did she eat? Write ratio as a reduced fraction.

6) Kristy read 20 pages. What part of the book did she read, if the book has 50 pages? Write ratio as a reduced fraction.

7) Lillian has 24 pencils. She gave her son 6 pencils. What part of the pencils did she give to her son? Write ratio as a reduced fraction.

8) Claire had \$36. She spent \$9. What part of her money did she spend? Write ratio as a reduced fraction.

9) Find the average for the numbers: $1\frac{1}{2}$, $2\frac{1}{5}$, $3\frac{1}{5}$

10) Find the average for the numbers: $4\frac{1}{2}$, $5\frac{1}{5}$, $6\frac{1}{5}$, $8\frac{1}{10}$

Grade 8 Chapter 4 Lesson 6

Word problems on Direct Proportionality
Directly Proportional Quantities

If first quantity increases or decreases, then the second quantity increases or decreases in the same way, then both quantities are called directly proportional.

a) If 4 apples sell for \$1.32. How much do 11 apples cost?

$132 \div 4 = 33$ c $\qquad$ $33 \times 11 = 363$ c = \$ 3.63

First Property of Direct Proportionality Quantities

If the ratio corresponding value of two quantities is a constant number then these quantities are directly proportional.

$132 \div 4 = 33$ c $\qquad$ $363 \div 11 = 33$ c

Second Property of Directly Proportional Quantities

If two quantities are directly proportional, then the ratios between corresponding value of two quantities is equal.

$132 \div 4 = 33$ c $\qquad$ $363 \div 11 = 33$ c

$132 \div 4 = 363 \div 11$

b) Pineapple costs \$12 for 4 lb. At this rate, how much would 3 lb of pineapple cost? $\quad 12 \div 4 = \$\,3 \qquad 3 \times 3 = \$\,9$

c) A car can travel 264 mi on 11 gallons of gas. At the same rate, how many gallons of gas would be needed to travel 528 mi?

$264 \div 11 = 24$ mi/gal $\qquad$ $528 \div 24 = 22$ gal

d) A candle will burn 2 in for 40 min. How many inches of the candle will burn in 1 hour? $\qquad$ 1 hour = 60 minutes

$40 \div 2 = 20$ min/in $\qquad$ $60 \div 20 = 3$ in

e) Ashley has lost 2 lb in 4 weeks. At this rate, how long will it take her to lose 5 lb?

$2 \div 4 = 0.5$ lb/week $\qquad$ $5 \div 0.5 = 10$ week

Grade 8 Chapter 4 Lesson 6
Review 1 A
Solve the Word Problems on Direct Proportionality

1) How many would 3 pens cost, if 5 costs $ 2.50?

2) How many would 5 notebooks cost, if 3 notebooks cost 12 c?

3) How many details can a worker do for 4 hours, if he does 28 details for 2 hours?

4) How many pillowcases can a woman sew for 6 days, if she can sew 12 pillowcases in 2 days?

5) A worker can make 15 details in 3 hours. How much time does he need to make 40 details?

6) A studio can sew 12 dresses in 3 days. How many days would they need if they want to sew 20 dresses?

7) Amy can type 18 pages in 6 hours. How many page can she type in 4 hours?

8) A carpenter can make 36 frames for windows in 3 days. How many frames can he make in 2 days?

9) Anna can bind 160 books in 20 days. How many books can she bind in 10 days?

10) A gardener can cut down 80 trees in 5 days. How many trees can he cut down in 3 days?

Grade 8 Chapter 4 Lesson 6
Review 1 B
Solve the Word Problems

1) A factory got 632 lb. of starch from 4 tons of potatoes.
How much starch can a factory get from 6 tons of potatoes?

2) A team made 750 keys in 3 hours. How many keys can the team make in 2 hours?

3) In a year, 10,000 m^2 of forest gives oxygen to 300 people.
How many people can get oxygen from 900,000m^2 of forest?

4) A factory got 16 lb. of apple puree from 20 lb. of apples.
How much apple puree can the factory get from 45 lb. of apples?

5) A factory got 5 lb. of oil from 20 lb. of seed.
How much oil can the factory get from 7 lb. of seed?

6) Brain paid $1.75 for 2.5 lb. of cookies.
How much of the same cookie can the boy buy for $2.45?

7) Linda paid $1.8 for 200 gram of candy.
How much does 1 kg of the same candy cost? 1 kg = 1000 g

8) David paid $60 for 15 balls. How many balls can he buy for $20?

9) A girl paid $20 for 4 books. How much 6 of the same would books cost?

10) Vicki paid $24 for 6 months for a newspaper. How many months can the woman get the same newspaper for $ 16?

Grade 8 Chapter 4 Lesson 7

Word Problems on Uniform Motion

The motion of a moving object whose speed and direction do not change is called uniform motion.

Word problem at moving content three quantities

Distance - d Time - t Rate – r (Speed)

Rate – Distance passed for time unit (for 1 hour, 1 minute)

Movement units

Distance (d) measure mi, m, km

Time (t) measure in min, hrs.

Rate (r) measure in mi/hrs., m/min, km/hrs.

Distance d

To find distance, multiply rate by time: $d = rt$

What distance did a man runs in 2 hours, if his rate was 5 mi/hr?

$d = 5 \times 2 = 10$ mi

If men drive at the speed of 70mph for 2 hours, how far will he travel? $d = 70\times2=140$ mi

Time t

To find time, divide distance by speed: $t = d \div r$

How much time needed if a man went 8 mi with a rate of 4 mi/hrs.?

$t = 8 \div 4 = 2$ hrs.

If a man drives at a speed of 52 mph for 91 mi, how long will he travel?

$t = d \div r = 91$ mi $\div$ 52 mph $= 7 \div 4 = 1.75$ hours

Rate (Speed) r

To find the speed, divide distance by time: $r = d \div t$

How much rate does a girl have, if she runs 6 mi in 2 hours.

$r = 6 \div 2 = 3$ mi / hr.

Woman drove a distance of 252 mi in $4\frac{2}{3}$ hours. What was her speed?

$r = d \div t = 252 \div 4\frac{2}{3} = \frac{252}{1} \times \frac{3}{14} = 18\times3 = 54$ mph

Grade 8 Chapter 4 Lesson 7
Review 1 A
Find the Unknowns of Uniform Motion

		Distance	Time	Speed
1) a	Pedestrian	x mi	4 hrs.	5 mi/hr.
1) b	Tractor	x mi	2 hrs.	30 mi/hr.
1) c	Train	x mi	3 hrs.	120 mi/hr.
1) d	Plane	x mi	2 hrs.	1000 mi/hr.

2) a	Skier	26 mi	x hrs.	13 mi/hr.
2) b	Train	240 mi	x hrs.	60 mi/hr.
2) c	Car	160 mi	x hrs.	80 mi/hr.
2) d	Plane	5000 mi	x hrs.	2500 mi/hr.

3) a	Bicycle	28 mi	2 hrs.	x mi/hr.
3) b	Car	210 mi	3 hrs.	x mi/hr.
3) c	Plane	15000 mi	3 hrs.	x mi/hr.

4) a	Eagle	x mi	1 hr.	120 mi/hr.
4) b	Starling	x mi	2 hrs.	80 mi/hr.
4) c	Swallow	x mi	3 hrs.	74 mi/hr.

5) a	Deer	x mi	3 hrs.	10 mi/hr.
5) b	Shark	72 mi	x hrs.	36 mi/hr.
5) c	Ship	280 mi	4 hrs.	x mi/hr.

6) a	Sparrow	120 mi	2 hrs.	x mi/hr.
6) b	Zebra	45 mi	x hrs.	15 mi/hr.
6) c	Boat	x mi	4 hrs.	24 mi/hr.

7) a	Car	The same	2 hrs.	60 mi/hr.
7) b	Car	The same	3 hrs.	x mi/hr.

Grade 8 Chapter 4 Lesson 7

Find the unknowns

	Distance	Time	Rate
8) a	x miles	2 hours	60 mi/hr.
8) b	120 miles	x hours	60 mi/hr.
8) c	120 miles	2 hours	x mi/hr.

9) a	x mi	2 hrs.	40 mi/hr.
9) b	24 mi	x hrs.	6 mi/hr.
9) c	300 mi	5 hrs.	x mi/hr.

10) a	70 m	7 secs	The same
10) b	90 m	x secs	The same

Review 1 B

1) A man runs with a speed of 5 mi/h and spent 4 hours. What distance did he pass?
2) A motorcyclist drove 2 hours with a speed of 60 km/h. What distance did he drive?
3) A boy runs 6 meters in 1 second. What distance can he run in 15 seconds?
4) A sound runs 330 meters in the air for 1 second. What distance can the sound run for 3 seconds?
5) Skier skies 3 hours with a speed of 12 km/h. What distance did he run?
6) A plane flies 675 km in one hour. What distance can the plane fly in 3 hours?
7) What distance does a boat travel in 7 hours, if it has a speed of 35 km/h?
8) A train travels 60 km in an hour. What distance does it travel in $\frac{3}{4}$ hour?
9) A bus drove 90 km at a speed of 45 km/h. How much time it spent?
10) What speed does a man have, if he runs 12 mi in 3 hours?

Grade 8 Chapter 4 Lesson 8

Word Problems on Percent

Converting Percent and Decimals, Fraction, or Mixed Number

Percent $\frac{1}{100} = 1\%$ **and** $0.01 = 1\ \%$

Converting a Decimal, Fraction or Mixed Number to a Percent

If a fraction, decimal or mixed number needs to convert to a percent, multiply the fraction, decimal or mixed number by 100%.

A decimal to a percent: $0.75 = 0.75 \times 100\% = 75\%$

A fraction to a percent: $\frac{3}{8} = \frac{3}{8} \cdot 100\% = \frac{75}{2}\% = 37\frac{1}{2}\% = 37.5\%$

A mixed number to a percent: $1\frac{4}{5} = 1.8 \times 100\% = 180\%$ $\quad 1 = 100\%$

If a percent needs to convert to a decimal, fraction, or mixed number divide the percent by 100%

To a decimal $50\% = \frac{50\%}{100\%} = 0.5$ **To a fraction** $25\% = \frac{25\%}{100\%} = \frac{1}{4}$

To a mixed number $150\% = \frac{150\%}{100\%} = \frac{3}{2} = 1\frac{1}{2}$ $\quad 100\% = 1$

Percent Proportion: $\frac{A}{B} = \frac{R}{100}$ $\quad 100A = BR$

In word problems in percent each word has special meanings "Amount" (A) represents R %, "Base" (B) represents 100%, "is" means =, "of'' means multiplication, "what number" means x

Percent Formulas: Amount: $A = \frac{BR}{100}$ $\quad A = B \cdot \frac{R}{100}$ (1)

Base: $B = \frac{100A}{R}$ $\quad B = A \cdot \frac{100}{R}$ $\quad B = A \div \frac{R}{100}$ (2)

Rate: $R = \frac{100A}{B}$ $\quad R = \frac{A}{B} \cdot 100\%$ (3)

If a percent is less than 100% then the Amount is less than the Base

$R < 100\%$ $\quad A < B$

If a percent is more than 100% then the Amount is greater than the Base: $R > 100\%$ $\quad A > B$

Grade 8 Chapter 4 Lesson 8

Solve Percent Problems by Equations ("is" and "of")

1) **Find A: What numbers is 25% of 40?** $A = 0.25 \cdot 40$ $\underline{A = 10}$

2) **Find B: 10 is 25% of what number?**

$10 = \frac{25}{100} B$ $\frac{10}{0.25} = \frac{0.25B}{0.25}$ $\frac{1000}{25} = B$ $\underline{B = 40}$

3) **Find R: What percent of 40 is 10?**

$\frac{R}{100} \times 40 = 10$ $100 \times \frac{R}{100} \times 40 = 10 \times 100$ $40R = 1000$ $\underline{R = 25\%}$

Solve Percent Problems by Proportion

1) **Find A: Michael solves 80% of 20 problems in test. How many problems he solved?**

$\frac{A}{B} = \frac{R}{100}$ $\frac{A}{20} = \frac{80}{100}$ $\frac{A}{20} = \frac{8}{10}$ $10A = 160$ $A = \underline{16 \text{ problems}}$

2) **Find B: David solved 16 problems in test, it is equal 80% of all test. How many problems were in test?**

$\frac{A}{B} = \frac{R}{100}$ $\frac{16}{B} = \frac{80}{100}$ $\frac{16}{B} = \frac{8}{10}$ $8B = 160$ $\underline{B = 20 \text{ problems}}$

3) **Find R: What percent of 20 is 16?**

$\frac{A}{B} = \frac{R}{100}$ $\frac{16}{20} = \frac{R}{100}$ $\frac{4}{5} = \frac{R}{100}$ $5R = 400$ $\underline{R = 80\%}$

Solve Percent Problems by Formulas

1) **Find A: Olivia had \$ 30, she spent 70% of all her money. How much money she spent? To find the amount, multiply the base by the percent, which converted to decimal.**

Formula 1 $\underline{A = B \cdot \frac{R}{100}}$ $A = 30 \cdot \frac{70}{100}$ $A = 30 \cdot 0.7$ $A = \underline{\$ 21}$

2) **Find B: Anna spent \$ 21, it is 70% of all her money. How much money she had? To find the base, divide the amount by the percent, which converted to decimal.**

Formula 2 $\underline{B = A \div \frac{R}{100}}$ $B = 21 \div \frac{70}{100}$ $B = 21 \div 0.7 = 210 \div 7 = \underline{\$ 30}$

3) **Find R: Vicki had \$ 30, she spent \$ 21. What percent of money she spent? To find a percent, divide the numbers and multiply the result by 100%. Formula 3**

$\underline{R = \frac{A}{B} \cdot 100\%}$ $R = \frac{21}{30} \cdot 100\%$ $R = \frac{7}{10} \times 100\%$ $R = \underline{70\%}$

To solve the percent problems, choose one of three ways that is easy for you.

Grade 8 Chapter 4 Lesson 8
Review 1 A
Solve the Word Problems on Percent

1) Find 20 % of 20?

2) Find 25 % of 200?

3) A book has 80 pages. The boy read 40% of the book. How many pages does he have left to read?

4) Of 40 children, 62.5% were boys. How many were girls?

5) The book has 240 pages. I read 40% of the book in the first day, then in the second day, I read 14 pages more than the first day. How many pages remain for me to read?

6) In the garden were 1720 trees: 25% of all trees were cherry and 15% of all trees were apple. How many cherry and apple trees total were in the garden?

7) Find the number if 20% of this number is 15?

8) Find the number, if 25% of this number is 6?

9) Find the number, if 50% of this number is 14?

10) Find the number, if 75% of this number is 30?

Grade 8 Chapter 4 Lesson 8

Review 1 A
Solve the Word Problems

11) On a test, Mike had 70% of the problems right. If he had 14 problems correct, how many questions were on the test?

12) A factory has 5 new machines and 80% were old machine. How many machines are in the factory?

13) From a box, Sam took 30 pages of paper. How many pages of paper was in the box, if in the box there remains 85% of all pages?

14) What percent is \$7 from \$8?

15) What percent is 1 lb from 8 lb?

16) What percent is 2 from 5?

17) What percent is 4 from 8?

18) In 50 minute of lesson, students solve an equation in 10 minutes. What percent of lesson time did students use to solve the equation?

19) Of 30 students, 6 students got an “A” on a test. What percent of students got an “A”?

20) There are 160 ml of acid in 800 ml of a solution of acid and water. What percent of the solution is acid?

Grade 8 Chapter 4 Lesson 8
Review 1 B
Solve the Word Problems

1) A bookstore manager determines that a markup rate of 30% is necessary to make a profit. What is the markup on a book that costs the bookstore \$ 10?

2) A disc player that regularly sells for \$ 400 is selling for 25% off the regular price. What is the discount?

3) Amy sold \$ 9000 of Jewelry this week. Find her commission for the week if she receives a commission rate of 5.6%?

4) How much commission will Jack make on the sale of a \$ 325,000 house if he receives 2% of the selling price?

5) A family paid \$ 25,000 as a down payment for a home. If this represents 20% of the price of the home, find the price of the home.

6) A student spent \$ 390 for books, if its 15% of cost for last semester at her community college. How much cost of last semester is at her community college?

7) The sales tax is \$ 96 on a stereo sound system purchase of \$ 1600. Find the sales tax rate.

8) The sales tax is \$ 108 on a necklace purchase of \$ 1200. Find the sales tax rate.

9) The college bookstore is giving a discount of \$8 on calculator that normally sell for \$ 32. What is the discount rate?

10) An electric grill that regularly sells for \$ 120 is selling for \$ 36 off the regular price. What is the discount rate?

Grade 8 Chapter 4 Lesson 8
Review 1 B
Solve the Word Problems

11) Amy was making a wage of $ 14.50 an hour before receiving a 12% in hourly pay. What Anna's new hourly wage?

12) A clothing store bought leather suit for $ 84 and used a markup rate of 45%. What is the selling price?

13) A $ 400 fax machine is on sale for 25% off. Find the amount of discount and the sale price.

14) A $ 4000 designer dress is on sale for 40% off. Find the amount of discount and the sale price.

15) The original price of a shirt is $ 60. What is a new price of a shirt with a 40% discount.

16) A lawn mower is on sale for 35% off the regular price of $ 1200. Find the sale price.

17) The whole price of coffee increased from $ 2 per pound to $ 3 per pound in one year. What was the percent increase in the price of 1-pound coffee?

18) The price of gasoline rose from $ 1.5 to $ 1.8 on five months. What was the percent increase in the price of gasoline?

19) A company with 1500 employees reduces the number of employees to 1200. What is the percent decrease?

20) The stereo television that regularly sells for $ 200 is on sale for $ 150. Find the discount rate.

Mental Math 17
Solve Equations

1) $x + 6 = 10$
2) $7 + x = 10$

3) $x - 9 = 5$
4) $x - 8 = 9$

5) $x - 7 = 8$
6) $12 - y = 3$

7) $13 - y = 8$
8) $14 - y = 5$

9) $15 - y = 7$
10) $16 - y = 9$

11) $8 \cdot x = 72$
12) $x \cdot 9 = 63$

13) $x \div 8 = 7$
14) $x \div 7 = 6$

15) $x \div 6 = 9$
16) $72 \div z = 9$

17) $72 \div z = 8$
18) $63 \div z = 9$

19) $56 \div z = 7$
20) $54 \div z = 9$

Grade 8 Chapter 4 Lesson 9

Mental Math 18

Table of Exponents 2

$a^1 = a$

1) 1^3	11) 2^1
2) 2^3	12) 2^2
3) 3^3	13) 2^3
4) 4^3	14) 2^4
5) 5^3	15) 2^5
6) 6^3	16) 2^6
7) 7^3	17) 2^7
8) 8^3	18) 2^8
9) 9^3	19) 2^9
10) 10^3	20) 2^{10}

Grade 8 Chapter 4 Lesson 9
Table 13 Word of Problems

1) Decimals

a) To find the amount by known base, multiply the base by a decimal:

Find the amount 0.6 of 40 $40 \times 0.6 = 24$

b) To find the base by known amount, divide the amount by a decimal: Finding the base: 0.6 is equivalent $ 24 $24 \div 0.6 = 40$

c) To find the ratio between two numbers, divide the numbers.

Find the ratio 24 from 40 $24 \div 40 = 0.6$

2) Fractions

a) To find the amount by known base, multiply the base by a fraction.

Finding the amount: $\frac{3}{5}$ of 40 $40 \times \frac{3}{5} = \$24$

b) To find the base by known amount, divide amount by a fraction.

Finding the base: $\frac{3}{5}$ equivalent 24 $24 \div \frac{3}{5} = \$40$

c) To find the ratio between two numbers, divide the numbers.

Find the ratio: 24 from $40 $24 \div 40 = \frac{24}{40} = \frac{3}{5}$

3) Uniform Motion: Distance - d Time - t Rate – r (Speed)

a) To find distance, multiply rate by time: $d = rt$

What distance did a man go for 2 hours if his rate was 5 mi/hr?

$d = 5 \cdot 2 = 10$ mi

b) To find time, divide distance by speed. $t = d \div r$

How much time does a man needs if he runs 8 mi with a rate 4 mi/hr?

$t = 8 \div 4 = 2$ hr

c) To find speed, divide distance by time. $r = d \div t$

What is the rate a girl has, if she runs 6 mi for 2 hours

$r = 6 \div 2 = 3$ mi / hr

4) Directly proportional quantities

If 4 apples sell for $1.32. How much do 11 apples cost?

$132 \div 4 = 33$ c $33 \times 11 = 363$ c = $3.63

Grade 8 Chapter 4 Lesson 9 Table 14 Percent Problems

Solve Percent Problems by Equations ("is" and "of")

1) Find A: What numbers is 25% of 40? $A = 0.25 \cdot 40$ $\underline{A = 10}$

2) Find B: 10 is 25% of what number?

$10 = \frac{25}{100}B \qquad \frac{10}{0.25} = \frac{0.25B}{0.25} \qquad \frac{1000}{25} = B \qquad \underline{B = 40}$

3) Find R: What percent of 40 is 10?

$\frac{R}{100} \times 40 = 10 \qquad 100 \times \frac{R}{100} \times 40 = 10 \times 100 \quad 40R = 1000 \quad \underline{R = 25\%}$

Solve Percent Problems by Proportion

1) Find A: Michael solves 80% of 20 problems in test. How many problems he solved?

$\frac{A}{B} = \frac{R}{100} \quad \frac{A}{20} = \frac{80}{100} \quad \frac{A}{20} = \frac{8}{10} \qquad 10A = 160 \qquad A = \underline{16 \text{ problems}}$

2) Find B: David solved 16 problems in test, it is equal 80% of all test. How many problems were in test?

$\frac{A}{B} = \frac{R}{100} \quad \frac{16}{B} = \frac{80}{100} \quad \frac{16}{B} = \frac{8}{10} \qquad 8B = 160 \qquad \underline{B = 20 \text{ problems}}$

3) Find R: What percent of 20 is 16?

$\frac{A}{B} = \frac{R}{100} \qquad \frac{16}{20} = \frac{R}{100} \quad \frac{4}{5} = \frac{R}{100} \qquad 5R = 400 \qquad \underline{R = 80\%}$

Solve Percent Problems by Formulas

1) Find A: Olivia had \$ 30, she spent 70% of all her money. How much money she spent? To find the amount, multiply the base by the percent, which converted to decimal.

Formula 1 $\underline{A = B \cdot \frac{R}{100}}$ $\quad A = 30 \cdot \frac{70}{100} \qquad A = 30 \cdot 0.7 \qquad A = \underline{\$ 21}$

2) Find B: Anna spent \$ 21, it is 70% of all her money. How much money she had? To find the base, divide the amount by the percent, which converted to decimal.

Formula 2 $\underline{B = A \div \frac{R}{100}}$ $\quad B = 21 \div \frac{70}{100} \qquad B = 21 \div 0.7 = 210 \div 7 = \underline{\$ 30}$

3) Find R: Vicki had \$ 30, she spent \$ 21. What percent of money she spent? To find a percent, divide the numbers and multiply the result by 100%. Formula 3

$\underline{R = \frac{A}{B} \cdot 100\%} \quad R = \frac{21}{30} \cdot 100\% \quad R = \frac{7}{10} \times 100\% \quad R = \underline{70\%}$

To solve the percent problems, choose one of three ways which is easy for you.

Grade 8 Chapter 4 Lesson 9
Table 15 Percent Application

1. Sales Tax Formula

Amount of sales tax = Cost of merchandise × Tax rate

Example: A graphing calculator cost \$ 100. The sales tax rate is 5.5%. Compute the total cost.

5.5% = 0.55 100×0.055 =\$5.5 100 + 5.5 = \$105.5

2. Percent of commission

Amount of commission = Dollar sales × Percent of commission

Example: Diana earned 5% on a house she sold for \$ 100,000

Find her dollar commission. 100000× 0.05 = \$ 5000

3. Percent of decrease

Amount of decrease = original value - new value

Amount of decrease = original value × percent of decrease

Percent of decrease = amount of decrease ÷ original value ×100%

Example: A paper shredder was decreased price from \$ 80 to \$ 60

What is the percent decreased in price?

80 – 60 = 20 20 ÷ 80 × 100% = 0.25 × 100% = 25%

4. Percent of discount

Amount of decrease = original value - new value

Amount of decrease = original value × percent of decrease

Percent of decrease = amount of decrease ÷ original value ×100%

Example: The Company put suits that usually sell for \$ 250 on sale for \$ 170.

What the percent discount?

250– 170 = \$ 80 80 ÷ 250 × 100% = 0.32 × 100% = 32%

5. Percent of increase

Amount of increase = new value – original value

Amount of increase = original value × percent of increase

Percent of increase = amount of increase ÷ original value × 100%

Example: The sales in one year Honda Civic increased from 400,000 to 408,000. Compute the percent increase.

408000 – 400000=8000 8000÷400000 ×100 = 0.2×100=20%

6. Percent markup

Amount of markup = new value – original value

Amount of markup = original value × percent of markup (rate)

Percent of markup = amount of markup ÷ original value × 100%

Example: A college bookstore marks up of math book is 30%.

What is mark up for math book that has a price of \$ 70.

\$70 × 0.3 = \$ 21 70 + 21 = \$ 91

Grade 8 Chapter 4 Lesson 9
Table 16 Common Factor
Distributive laws of multiplication

1) Whole Numbers (Put A before parentheses)

$AB + AC = A(B + C)$ $AB - AC = A(B - C)$

If two products have a common factor, put the common factor before parentheses.

$AB + AC$ A – common factor

In parentheses, put results of division products by common factor

$AB \div A = B$ $AC \div A = C$

$16 \cdot 5 + 4 \cdot 5 = 5(16 + 4) = 5(20) = 100$

$22 \cdot 5 - 2 \cdot 5 = 5(22 - 2) = 5(20) = 100$

2) Decimals

a) $(B + C)A = AB + AC$ $(0.4 + 0.1)5 = 0.4 \times 5 + 0.1 \times 5 = 2 + 0.5 = 2.5$

$AB + AC = A(B + C)$ $0.4 \times 5 + 0.1 \times 5 = 5(0.4 + 0.1) = 5(0.5) = 2.5$

b) $A(B - C) = AB - AC$ $5(0.4 - 0.1) = 5 \times 0.4 - 5 \times 0.1 = 2 - 0.5 = 1.5$

$AB - AC = A(B - C)$ $5 \times 0.4 - 5 \times 0.1 = 5(0.4 - 0.1) = 5(0.3) = 1.5$

3) Negative Numbers

If first term negative, put (- A) before parentheses and change each sign inside parentheses: $-AB - AC = -A(B + C)$ $-AB + AC = -A(B - C)$

a) $-16 \cdot 5 - 4 \cdot 5 = -5(16 + 4) = -5(20) = -100$

$-22 \cdot 5 + 2 \cdot 5 = -5(22 - 2) = -5(20) = -100$

b) $-7.5 \times 5 - 2.5 \times 5 = -5(7.5 + 2.5) = -5(10) = -50$

$-12.5(5) + 2.5(5) = -5(12.5 - 2.5) = -5(10) = -50$

4) To solve problems work by order of operations or change division for multiplication then put A before parentheses.

$AB + AC = A(B + C)$ $AB - AC = A(B - C)$

a) $2\frac{5}{8} \div \frac{1}{8} + 1\frac{3}{8} \div \frac{1}{8} = 2\frac{5}{8} \times 8 + 1\frac{3}{8} \times 8 = 8(2\frac{5}{8} + 1\frac{3}{8}) = 8(4) = 32$

b) $2\frac{5}{8} \div \frac{1}{8} - 1\frac{3}{8} \div \frac{1}{8} = 2\frac{5}{8} \times 8 - 1\frac{3}{8} \times 8 = 8(2\frac{5}{8} - 1\frac{3}{8}) = 8(1\frac{2}{8}) = 8(\frac{10}{8}) = 10$

5) To solve problems work by order of operations or change division for multiplication then put (- A) before parentheses

$-AB - AC = -A(B + C)$ $-AB + AC = -A(B - C)$

a) $-2\frac{5}{8} \div \frac{1}{8} - 1\frac{3}{8} \div \frac{1}{8} = -2\frac{5}{8} \times 8 - 1\frac{3}{8} \times 8 = -8(2\frac{5}{8} + 1\frac{3}{8}) = -8(4) = -32$

b) $-2\frac{5}{8} \div \frac{1}{8} + 1\frac{3}{8} \div \frac{1}{8} = -2\frac{5}{8} \times 8 + 1\frac{3}{8} \times 8 = -8(2\frac{5}{8} - 1\frac{3}{8}) = -8(1\frac{2}{8}) = -8(\frac{10}{8}) = -10$

Grade 8 Chapter 4 Lesson 10
Laws for operations
Distributive laws of multiplication

(A + B) C = AC + BC **(A - B) C = AC – B C**

Example a) $\frac{5}{6}\times 3=\frac{5}{6}\times\frac{3}{1}=\frac{5}{2}=2\frac{1}{2}=2.5$

Example b) $\frac{5}{3}\times 6=6\times\frac{5}{3}=6\div 3\times 5=2\times 5=10$

Example c) $\left(1\frac{1}{3}-\frac{5}{6}\right)\times 3=\frac{4}{3}\times\frac{3}{1}-\frac{5}{6}\times\frac{3}{1}=4-\frac{5}{2}=4-2.5=1.5$

Worksheet 6 A
Evaluate by using laws

1) $9(1-\frac{4}{9})$

2) $(\frac{2}{3}-\frac{3}{5})\times 15$

3) $9(3-\frac{5}{9})$

4) $(1+\frac{5}{6})\times 6$

5) $7(2+\frac{3}{7})$

6) $(1-\frac{1}{6})\times 6$

7) $6(1\frac{1}{2}-\frac{2}{3})$

8) $(\frac{5}{7}+\frac{4}{7})\times 7$

9) $7(\frac{5}{7}-\frac{4}{7})$

10) $(3+\frac{2}{3})\times 3$

11) $3(2+\frac{1}{6})$

12) $(2\frac{1}{4}-\frac{5}{8})\times 4$

13) $3(1\frac{5}{6}-1\frac{1}{3})$

14) $(3\frac{1}{2}-\frac{1}{4})\times 4$

15) $3(1\frac{2}{3}-\frac{5}{6})$

16) $(1\frac{2}{7}-\frac{1}{3})\times 21$

17) $6(2\frac{1}{2}+1\frac{1}{3})$

18) $(4+\frac{5}{8})\times 2$

19) $3(1\frac{1}{3}-\frac{5}{12})$

20) $(1\frac{1}{4}-\frac{5}{8})\times 4$

Grade 8 Chapter 4 Lesson 10
Worksheet 6 B
Distributive laws of multiplication
Common Factor (Put C before parentheses)

AC + BC = C (A + B) **AC – B C = C (A - B)**

Example $1\frac{1}{4}\times\frac{3}{4}+\frac{3}{4}\times2\frac{3}{4}=\frac{3}{4}(1\frac{1}{4}+2\frac{3}{4})=\frac{3}{4}\times\frac{4}{1}=3$

Evaluate by using laws

1) $8\times1\frac{3}{4}-8\times\frac{5}{8}$

2) $3\frac{2}{9}\times\frac{3}{5}+\frac{3}{5}\times6\frac{7}{9}$

3) $7\times1\frac{1}{7}-7\times\frac{5}{14}$

4) $4\times2\frac{1}{4}-4\times1\frac{7}{8}$

5) $2\frac{6}{7}\times\frac{7}{10}+2\frac{1}{7}\times\frac{7}{10}$

6) $5\frac{5}{6}\times\frac{3}{4}-1\frac{5}{6}\times\frac{3}{4}$

7) $4\frac{4}{5}\times\frac{5}{8}+\frac{5}{8}\times3\frac{1}{5}$

8) $3\frac{2}{3}\times1\frac{1}{2}-1\frac{2}{3}\times1\frac{1}{2}$

9) $\frac{5}{6}\times5\frac{1}{5}-\frac{5}{6}\times2\frac{1}{5}$

10) $3\frac{1}{3}\times1\frac{1}{5}+1\frac{1}{5}\times1\frac{2}{3}$

11) $1\frac{3}{5}\times2\frac{1}{2}+1\frac{3}{5}\times7\frac{1}{2}$

12) $4\frac{2}{3}\times\frac{3}{4}-3\frac{1}{3}\times\frac{3}{4}$

13) $4\frac{3}{5}\times\frac{5}{6}-\frac{5}{6}\times1\frac{3}{5}$

14) $4\frac{1}{2}\times\frac{1}{6}+\frac{1}{6}\times1\frac{1}{2}$

15) $3\frac{3}{4}\times1\frac{1}{6}-2\frac{1}{4}\times1\frac{1}{6}$

16) $\frac{1}{3}\times15-\frac{1}{5}\times15$

17) $1\frac{3}{4}\times8-1\frac{1}{2}\times8$

18) $1\frac{1}{2}\times6-1\frac{1}{6}\times6$

19) $2\frac{1}{2}\times6-1\frac{1}{2}\times6$

20) $1\frac{3}{4}\times12-1\frac{1}{4}\times12$

21) $1\frac{3}{4}+1\frac{1}{4})\div\frac{1}{4}$

22) $(1\frac{3}{5}-1\frac{2}{5})\div\frac{2}{5}$

23) $(4\frac{3}{5}+1\frac{2}{5})\div\frac{1}{5}$

24) $(1\frac{3}{4}-1\frac{1}{4})\div\frac{1}{8}$

25) $(2\frac{1}{2}+1\frac{1}{2})\div\frac{1}{6}$

26) $(3\frac{2}{3}-1\frac{2}{3})\div\frac{2}{3}$

27) $(1\frac{5}{6}+1\frac{1}{6})\div\frac{1}{6}$

28) $(4\frac{3}{5}-1\frac{3}{5})\div\frac{1}{5}$

29) $(1\frac{2}{3}+1\frac{1}{3})\div\frac{1}{3}$

30) $(4\frac{5}{6}-2\frac{1}{6})\div\frac{1}{6}$

Grade 8 Chapter 4 Lesson 11
Reviews for Chapter 4
Review 4 A

Date **Name**
Read directions carefully. **Answers must be in the indicated place.**
Show all steps for full credits. **Do only 1 operation per 1 step.**
Each problem is worth 5 points. **Solve 1 problem 5 minutes.**

1) A book has 80 pages. David read 0.15 of the whole book.
How many pages more does he have left to read?

2) In the class, 5 students can dance; this is $\frac{1}{4}$ of the whole class. How many students are in the class? How many students cannot dance?

3) Kristy read 10 pages. What part of the book did she read, if the book has 50 pages? Write the ratio as a reduced fraction. How many pages more does she have left to read?

4) Find the average for the numbers: $\frac{1}{2}, \frac{1}{5}, \frac{1}{5}, \frac{1}{10}$

5) A girl paid \$24 for 2 books. How much would 3 of the same books cost?

6) A bus drive 42 mi at a speed of 15 mi/h.
How much time it spent?

7) What speed does a boy have, if he drove 45 mi in 4 hours?

8) Of 40 children, 37.5% were boys. How many were girls?

9) A factory has 2 new machines and 80% were old machine.
How many machines are in the factory?

10) In 50 minutes of lesson, students solve an equation in 10 minutes.
What percent of lesson time did students use to solve the equation?

Grade 8 Chapter 4 Lesson 11
Review 4 B

Date
Name
Read directions carefully.
Answers must be in the indicated place.
Show all steps for full credits.
Do only 1 operation per 1 step.
Each problem is worth 5 points.
Solve 1 problem 5 minutes.

1) A book has 60 pages. David read 0.55 of the whole book. How many pages more does he have left to read?

2) In the class, 8 students can dance; this is $\frac{2}{5}$ of the whole class. How many students are in the class? How many students cannot dance?

3) Kristy read 8 pages. What part of the book did she read, if the book has 40 pages? Write the ratio as a reduced fraction. How many pages more does she have left to read?

4) Find the average for the numbers: $1\frac{1}{2}$, $1\frac{1}{5}$, $3\frac{1}{5}$, $3\frac{1}{10}$

5) A girl paid $12 for 2 books. How much would 3 of the same books cost ?

6) A bus drove 36 mi at a speed of 15 mi/h. How much time it spent?

7) What speed does a boy have, if he drove 61 mi in 4 hours?

8) Of 24 children, 37.5% were boys. How many were girls?

9) A factory has 4 new machines and 60% were old machine. How many machines are in the factory?

10) In 75 minutes of lesson, students solve an equation in 15 minutes. What percent of lesson time did students use to solve the equation?

Grade 8 Chapter 4 Lesson 12
Tests for Chapter 4
Test 4 A

Date | **Name**
Read directions carefully. | **Answers must be in the indicated place.**
Show all steps for full credits. | **Do only 1 operation per 1 step.**
Each problem is worth 5 points. | **Solve 1 problem 5 minutes.**

1) A book has 80 pages. David read 0.25 of the whole book. How many pages more does he have left to read?

1) ______

2) In the class, 6 students can dance; this is $\frac{1}{5}$ of the whole class. How many students are in the class? How many students cannot dance?

2) ______

3) Kristy read 20 pages. What part of the book did she read, if the book has 50 pages? Write the ratio as a reduced fraction. How many pages more does she have left to read?

3) ______

4) Find the average for the numbers: $\frac{1}{2}$, $\frac{1}{5}$, $1\frac{1}{5}$, $1\frac{1}{10}$

4) _______

5) A girl paid $15 for 3 books. How much would 4 of the same books cost?

5) _______

Grade 8 Chapter 4 Lesson 12

Test 4 A

Date **Name**

Read directions carefully. **Answers must be in the indicated place.**

Show all steps for full credits. **Do only 1 operation per 1 step.**

Each problem is worth 5 points. **Solve 1 problem 5 minutes.**

6) A bus drove 54 mi at a speed of 20 mi/h.
How much time it spent?

6) ______

7) What speed does a boy have, if he drove 49 mi in 4 hours?

7) ______

8) Of 40 children, 62.5% were boys. How many were girls?

8) ______

9) A factory has 3 new machines and 80% were old machine.
How many machines are in the factory?

9) ______

10) In 50 minutes of lesson, students solve an equation in 15 minutes.
What percent of lesson time did students use to solve the equation?

10) ______

Grade 8 Chapter 4 Lesson 12
Test 4 B

Date **Name**
Read directions carefully. **Answers must be in the indicated place.**
Show all steps for full credits. **Do only 1 operation per 1 step.**
Each problem is worth 5 points. **Solve 1 problem 5 minutes.**

1) A book has 60 pages. David read 0.65 of the whole book.
How many pages more does he have left to read?

1) _____

2) In the class, 9 students can dance; this is $\frac{3}{5}$ of the whole class. How many students are in the class? How many students cannot dance?

2) ______

3) Kristy read 16 pages. What part of the book did she read, if the book has 40 pages? Write the ratio as a reduced fraction. How many pages more does she have left to read?

3) ______

4) Find the average for the numbers: $3\frac{1}{2}$, $3\frac{1}{5}$, $1\frac{1}{5}$, $1\frac{1}{10}$

4) ______

5) A girl paid $12 for 3 books. How much would 4 of the same books cost?

5) ______

Grade 8 Chapter 4 Lesson 12
Test 4 B

Date **Name**

Read directions carefully. **Answers must be in the indicated place.**
Show all steps for full credits. **Do only 1 operation per 1 step.**
Each problem is worth 5 points. **Solve 1 problem 5 minutes.**

6) A bus drove 46 mi at a speed of 20 mi/h. How much time it spent?

6) ______

7) What speed does a boy have, if he drove 65 mi in 4 hours?

7) ______

8) Of 24 children, 62.5% were boys. How many were girls?

8) ______

9) A factory has 6 new machines and 60% were old machine. How many machines are in the factory?

9) ______

10) In 75 minutes of lesson, students solve an equation in 30 minutes. What percent of lesson time did students use to solve the equation?

10) _____

Grade 8 Answers Chapter 4

Lesson 1 1) 2.3 hr 2) 16.25 mi/hr 3) 9 girls 4) 15 machines 5) 40%

Lesson 2 Word problems on Operations

Review 1 A

1) 59 flowers 2) 89 roses 3) 69 cents 4) 58 birds 5) 88 cents
6) 2 pencils 7) 16 balloons 8) 11 roses 9) 24 hammers 10) 40 cents

Review 1 B

1) 80 hours 2) 280 seats 3) 225 words 4) 120 seats 5) 40 hours
6) 9 packages 7) 3 hours 8) 4 floors 9) 7 people 10) 9 students

Review 1 C

1) 46 markers 2) 69 nickels 3) 5 pens 4) 22 cups 5) 50 seats
6) 50 players 7) 60 sheets 8) 6 trees 9) 4 hours 10) 5 pages

Lesson 3 Word problems on Worth

Review 1 A

1) 10 c 2) 20 c 3) 30 c 4) 90 c 5) 72 c
6) $ 60 7) $ 50 8) 20 plants 9) 8 liters 10) 12 lb

Review 1 B

1) 8 buttons 2) 20 meters 3) $ 240 4) 90 c 5) $ 12
6) $ 4 7) 7 c 8) 20 envelopes 9) $ 9 10) 4 times

Review 1 C

1) 7 books 2) 6 tickets 3) 3 boys 4) 6 pencils 5) 3 bouquets
6) 8 logs 7) 2 bunches 8) $ 24 9) 4 packs 10) $ 280

Lesson 4 Word Problems on Decimals

Review 1 A

1) 7 students 2) 30 pages 3) 3 stamps 4) 4 dresses 5) 9 girls
6) 9pages 7) $ 60 8) 10 plums 9) 25 pages 10) 32 students

Review 1 B

1) 0.5 2) 0.4 3) 0.25 4) 0.25 5) 540 lb
6) 60 pages 7) 15 boys 8) $ 1.5 9) 2.3 10) 6

Lesson 5 Word Problems on Fractions

Review 1 A

1) 7 students 2) 30 pages 3) 3 stamps 4) 4 dresses 5) 9 girls
6) 9 pages 7) 540 lb 8) 60 pages 9) 15 girls 10) $ 1.5

Review 1 B

1) $ 60 2) 10 plums 3) 100 pages 4) 40students 5) $\frac{1}{2}$
6) $\frac{2}{5}$ 7) $\frac{1}{4}$ 8) $\frac{1}{4}$ 9) 2.3 10) 6

Grade 8 Answers Chapter 4

Lesson 6 Word Problems on Direct Proportionality

Review 1 A

1) \$1.50 2) 20 c 3) 56 details 4) 36 cases 5) 8 hr
6) 5 days 7) 12 pages 8) 24 frames 9) 80 books 10) 48 trees

Review 1 B

1) 948 lb 2) 500 keys 3) 27000 people 4) 36 lb 5) 1.75 lb
6) 3.5lb 7) \$9 8) 5 balls 9) \$30 10) 4 mo

Lesson 7 Uniform Motion

Review 1 A

1a) 20 mi 1b) 60 mi 1c) 360 mi 1d) 2000 mi

2a) 2 hrs. 2b) 4 hrs. 2c) 2 hrs. 2d) 2 hrs.

3a) 14 mi/hr. 3b) 70 mi/hr. 3c) 5,000 mi/hr.

4a) 120 mi 4b) 160 mi 4c) 222 mi

5a) 30 mi 5b) 2 hrs. 5c) 70 mi/hr.

6a) 60 mi/hr. 6b) 3 hrs. 6c) 96 mi

7a) 120 mi 7b) 40 mi/hr.

8a) 120 mi 8b) 2 hrs. 8c) 60 mi/hr.

9a) 80 mi 9b) 4 hrs. 9c) 60 mi/hr.

10a) 10 m/sec. 10b) 9 secs.

Review 1 B

1) 20 mi 2) 120 km 3) 90 m 4) 990 m 5) 36 mi

6) 2,025 mi 7) 245 mi 8) 45 mi 9) 2 hrs. 10) 4 mi/hr.

Grade 8 Answers Chapter 4

Lesson 8 Word Problems on Percent

Review 1 A

1) 4 2) 50 3) 60 pages 4) 15 girls 5) 34 pages
6) 688 trees 7) 75 8) 24 9) 28 10) 40
11) 20 questions 12) 25 machines 13) 200 pages 14) 87.5% 15) 12.5%
16) 40% 17) 50% 18) 20% 19) 20% 20) 20%

Review 1 B

1) $A = 10\times0.3 = \$\,30$ 2) $A = 400\times0.25 = \$100$

3) $A = 9000 \times 0.056 = \$\,504$ 4) $A=325{,}000\times 0.02= 6{,}500$

5) $B = 25000 \div 0.2 = \$\,125{,}000$ 6) $B = 390 \div 0.15 = \$2{,}600$

7) $R = 96 \div 1600 \times 100\% = 6\ \%$ 8) $R = 108 \div 1200 \times 100\% = 9\%$

9) $R = 8 \div 32 \times 100\% = 25\%$ 10) $R = 36 \div 120 \times 100\% = 30\%$

11) $A = 14.50 \times 0.12 = \$\,1.74$ $14.50 + 1.74 = \$\,16.24$

12) $A = 84 \times 0.45 = \$\,37.8$ $84 + 37.8 = \$\,121.8$

13) $A = 400 \times 0.25 = \$\,100$ $400 - 100 = \$\,300$

14) $A = 4000 \times 0.25 = \$\,1{,}000$ $4000 - 1000 = \$\,3{,}000$

15) $A = 60 \times 0.4 = \$\,24$ $60 - 24 = \$\,36$

16) $A = 1200 \times 0.35 = \$\,420$ $1200 - 420 = \$\,780$

17) $3 - 2 = 1$ $R = \frac{1}{2} \times 100\% = 50\%$

18) $1.8 - 1.5 = 0.3$ $R = 0.3 \div 1.5 \times 100\% = 20\%$

19) $1500 - 1200 = 300$ $R = 300 \div 1500 \times 100\% = 20\%$

20) $200 - 150 = 50$ $R = 50 \div 200 \times 100\% = 25\%$

Grade 8 Answers Chapter 4

Lesson 9 Mental Math 17

1) 4	2) 3	3) 14	4) 17	5) 15
6) 9	7) 5	8) 9	9) 8	10) 7
11) 9	12) 7	13) 56	14) 42	15) 54
16) 8	17) 9	18) 7	19) 8	20) 6

Lesson 9 Mental Math 18

1) 1	2) 8	3) 27	4) 64	5) 125
6) 216	7) 343	8) 512	9) 729	10) 1000
11) 2	12) 4	13) 8	14) 16	15) 32
16) 64	17) 128	18) 256	19) 512	20) 1024

Lesson 10 Worksheet 6 A

1) 5	2) 1	3) 22	4) 11	5) 17
6) 5	7) 5	8) 9	9) 1	10) 11
11) 6.5	12) 6.5	13) 1.5	14) 13	15) 2.5
16) 20	17) 23	18) 9.25	19) 2.75	20) 2.5

Lesson 10 Worksheet 6 B

1) 9	2) 6	3) 5.5	4) 1.5	5) 3.5
6) 3	7) 5	8) 3	9) 2.5	10) 6
11) 16	12) 1	13) 2.5	14) 1	15) 1.75
16) 2	17) 2	18) 2	19) 6	20) 6
21) 12	22) 0.5	23) 30	24) 4	25) 24
26) 3	27) 18	28) 15	29) 9	30) 16

Lesson 11 Review 4 A

1) 68 pages 2) 15 students 3) $\frac{1}{5}$, 40 pages 4) 0.25 5) \$ 36

6) 2.8 hr 7) 11.25 mi/hr 8) 25 girls 9) 10 machines 10) 20%

Lesson 11 Review 4 B

1) 27 pages 2) 12 students 3) $\frac{1}{5}$, 32 pages 4) 2.25 5) \$ 18

6) 2.4 hr 7) 15.25 mi/hr 8) 15 girls 9) 10 machines 10) 20%

Lesson 12 Test 4 A

1) 60 pages 2) 24 students 3) $\frac{2}{5}$, 30 pages 4) 0.75 5) \$20

6) 2.7 hr 7) 12.25 mi/hr 8) 15 girls 9) 15 machines 10) 30%

Lesson 12 Test 4 B

1) 21 pages 2) 6 students 3) $\frac{2}{5}$, 24 pages 4) 2.25 5) \$16

6) 2.3 hr 7) 16.25 mi/hr 8) 9 girls 9) 15 machines 10) 40%

Grade 8

Chapter 5

Summary of

Basic Math

Grade 8 Chapter 5 Contents
Summary of Basic Math
Lesson 1 – 14
Basic Problems

Lesson 1 Work with Tables 1, 2, 3, 4, 5, 6, 7

Lesson 2 Work with Tables 8, 9, 10, 11, 12, 13

Lesson 3 Work with Tables 14, 15, 16, 17, 18, 19

Lesson 4 Work with Tables 20, 21, 22, 23, 24, 25

Lesson 5 Work with Tables 26, 27, 28, 29, 30, 31

Teacher can put Mental Math in plastic for work in class (Lesson 6 – 9)

Lesson 6 Work with Mental Math 1, 2, 3, 4, 5

Lesson 7 Work with Mental Math 6, 7, 8, 9, 10, 11

Lesson 8 Work with Mental Math 12, 13, 14, 15, 16

Lesson 9 Work with Mental Math 17, 18, 19, 20, 21

Lesson 10 Review for Final Exam 1-A, B, C, D

Lesson 11 Worksheet 7 A, Worksheet 7 B

Lesson 12 Worksheet 8 A, Worksheet 8 B

Lesson 13 <u>College Placement Test A # 1 - # 10 (Basic Problems)</u>
<u>College Placement Test B # 1 - # 10 (Basic Problems)</u>

Lesson 14 <u>Mini Final Exam A # 1 – # 10 (Basic Problems)</u>
<u>Mini Final Exam B # 1 – # 10 (Basic Problems)</u>

Answers Chapter 5

Grade 8 Chapter 5 Lesson 1

Table 1
Addition and Subtraction of Whole Numbers

All numbers are made up of the digits: 0, 1, 2, 3, 4, 5, 6, 7, 8, and 9.
Basic operations are: addition, subtraction, multiplication, and division.
Natural (counting) numbers are numbers that use for calculations 1, 2, 3 …
Whole numbers are natural numbers and zero: 0, 1, 2, 3, 4, 5 …
Whole numbers can have represented with points on a number line

The first operation with numbers is addition, which combines two or more numbers together.

The first number in addition is called the addend	2
The second number in addition is called the addend	3
The result of addition is called the sum	5
Addend + Addend = Sum	2 + 3 = 5

The second operation with numbers is subtraction, which means to take something away.

The first number in subtraction is called the minuend	5
The second number in subtraction is called the subtrahend	2
The result of subtraction is called the difference	3
Minuend – Subtrahend = Difference	5 – 2 = 3

To check addition use subtraction

200 + 300 = 500	500 – 200 = 300	500 – 300 = 200

To check subtraction use addition or subtraction.

500 – 200 = 300	500 – 300 = 200	200 + 300 = 500

Remember

10 = 9 + 1 = 8 + 2 = 7 + 3 = 6 + 4 = 5 + 5	15 = 9 + 6 = 8 + 7
11 = 9 + 2 = 8 + 3 = 7 + 4 = 6 + 5	16 = 9 + 7 = 8 + 8
12 = 9 + 3 = 8 + 4 = 7 + 5 = 6 + 6	17 = 9 + 8
13 = 9 + 4 = 8 + 5 = 7 + 6	18 = 9 +9
14 = 9 + 5 = 8 + 6 = 7 + 7	19 = 10 + 9

Grade 8 Chapter 5 Lesson 1

Table 2

Multiplication and Division of Whole Numbers

Multiplication of Whole Numbers

The third operation with numbers is multiplication $\times$ or $\cdot$

Multiplication is the operation of repeated addition of the same numbers

$2\times 3 = 2 + 2 + 2 = 6$ or $2 \cdot 3 = 2 + 2 + 2 = 6$

The first number in multiplication is called the factor 2

The second number in multiplication is called the factor 3

The result of multiplication is called the product 6

Factor × Factor = Product $2 \times 3 = 6$

Multiplying Numbers by 10, 100 ...

To multiply numbers by 10, add zero to the end of a number: $5 \times 10 = 50$

To multiply numbers by 100, add two zeros to end of a number: $5 \times 100 = 500$

Division of Whole Numbers

The fourth operation with numbers is division $\div$ or :

Division means to separated number by some equal groups

The first number in division is called the dividend 6

The second number in division is called the divisor 2

The result of division is called the quotient 3

Dividend ÷ Divisor = Quotient $6\div2 = 3$

Division by 10

To divide numbers by 10, delete one zero: $40 \div 10 = 4$

Division by 100

To divide numbers by 100, delete two zeros: $400 \div 100 = 4$

To check multiplication use division

$2\times3=6$ Check $6\div2=3$ or $6\div3=2$

To check division use multiplication or division

$6\div2=3$ Check $2\times3=6$ or $6\div3=2$

Even numbers 16, 48

Numbers where last digits are even 0, 2, 4, 6, 8 called even numbers

Odd numbers 31, 27

Number where last digits are odd 1, 3, 5, 7, 9, called odd numbers

Grade 8 Chapter 5 Lesson 1

Table 3

Multiplication and Division

A + 0 = A	0 + A = A	A – 0 = A	A – A = 0
A × 1 = A	1 × A = A	A÷ 1 = A	A÷ A = 1
A × 0 = 0	0 × A = 0	0 ÷ A = 0	A ÷ 0 = ∅
A + B = B + A		A× B = B × A	∅ no solutions

Number 5 from the left column multiply by number 4 from first row to get answer 20. 5×4 = 20 20÷5=4 20÷4=5

Memorize Table

×	1	2	3	<u>4</u>	5	6	7	8	9	10
1	1	2	3	4	5	6	7	8	9	10
2	2	4	6	8	10	12	14	16	18	20
3	3	6	9	12	15	18	21	24	27	30
4	4	8	12	16	20	24	28	32	36	40
<u>5</u>	5	10	15	<u>20</u>	25	30	35	40	45	50
6	6	12	18	24	30	36	42	48	54	60
7	7	14	21	28	35	42	49	56	63	70
8	8	16	24	32	40	48	56	64	72	80
9	9	18	27	36	45	54	63	72	81	90
10	10	20	30	40	50	60	70	80	90	100

Grade 8 Chapter 5 Lesson 1

Table 4

Perfect Square Numbers

Take number 20 from the left column add number 5 from the first row to get number 25. Answer is in intersection row of 20 and column of 5 $25^2 = 25 \cdot 25 = 625$

Memorize Table of Perfect Square for Indicated Numbers

a^2	0	1	2	3	4	5	6	7	8	9
	0	1	4	9	16	25	36	49	64	81
10	100	121	144	169	196	225	256	289	324	361
20	400	441	484	529	576	625	676	729	784	841
30	900	961	1024	1089	1156	1225	1296	1369	1444	1521
40	1600	1681	1764	1849	1936	2025	2116	2209	2304	2401
50	2500	2601	2704	2809	2916	3025	3136	3249	3364	3481
60	3600	3721	3844	3969	4096	4225	4356	4489	4624	4761
70	4900	5041	5184	5329	5476	5625	5776	5929	6084	6241
80	6400	6561	6724	6889	7056	7225	7396	7569	7744	7921
90	8100	8281	8464	8649	8836	9025	9216	9409	9604	9801

Square Roots from Perfect Square Numbers

The square root of three and four-digit perfect square numbers is always a two-digit number. Divide the number under the root by groups of two digits, from right to left. The last group can be one or two digits. To find the first digit, find the number whose square is smaller than or equal to the last group. To find the second digit, use the rule:

Last digit of given number	1	4	5	6	9
Last digit of answer	1 or 9	2 or 8	5	4 or 6	3 or 7

$31^2 = 31 \times 31 = 961$ $32^2 = 32 \times 32 = 1024$ $\sqrt{9'61} = 31$ $\sqrt{10'24} = 32$

Grade 8 Chapter 5 Lesson 1

Table 5

Power of Whole Numbers.

The four arithmetic operations with numbers are: Addition, Subtraction, Multiplication, Division, $3 + 2 = 5$ $5 - 2 = 3$ $3 \times 2 = 6$ $6 \div 2 = 3$

The fifth operation with numbers is power of numbers.

A power of a number indicates repeated multiplication.

Base is a number that has to be multiplied 3

Exponent is a number that shows how many times to multiply the base 2

Power is result of multiplication of the same numbers 9

$3^2 = 3 \times 3 = 9$ Base ^{Exponent} = Power

Exponent 2 called squared $4^2 = 4 \times 4 = 16$

Exponent 3 called cubed $2^3 = 2 \cdot 2 \cdot 2 = 4 \cdot 2 = 8$

Special powers $0^{10} = 0$ $0^n = 0$ $1^{10} = 1$ $1^n = 1$ $a^1 = a$

Memorize perfect square numbers: 1, 4, 9, 16, 25, 36, 49, 64, 81, 100

$10^2 + 11^2 + 12^2 = 13^2 + 14^2$ $100 + 121 + 144 = 169 + 196$ $365 = 365$

$3^4 = 3 \times 3 \times 3 \times 3 = 81$ $3^5 = 81 \times 3 = 243$ $3^6 = 243 \times 3 = 729$

$4^4 = 4 \times 4 \times 4 \times 4 = 256$ $4^5 = 256 \times 4 = 1024$ $5^4 = 125 \times 5 = 625$

Prime Factorization

Factor × Factor = Product $3 \times 2 = 6$

A factor of a number n is a nonzero whole number that divides evenly into n

All divisors of 12 are 1, 2, 3, 4, 6, 12 $1 \times 12 = 2 \times 6 = 3 \times 4 = 12$

All divisors of 24 are 1, 2, 3, 4, 6, 8, 12, 24 $1 \times 24 = 2 \times 12 = 3 \times 8 = 4 \times 6 = 24$

Prime Numbers: A natural number called a prime number if it has only two different divisors: one and the number itself 2, 3, 5, 7, 11, 13, 17, 19 …

Composite Numbers: A natural number that has more than two divisors called a composite number 4, 6, 8, 9, 10, 12, 14, 15, 16…

The number 1 has only one divisor: which is why 1 is not a prime or a composite.

Prime numbers cannot factorize. Remember the first 10 prime numbers: 2, 3, 5, 7, 11, 13, 17, 19, 23, 29

To factorize composite numbers to prime factors means to represent composite numbers in the product of prime factors: $8 = 4 \times 2 = 2 \times 2 \times 2 = 2^3$

Prime Factorization

Example 1: Find prime factorization for 100

Try to divide by prime numbers: by 2, by 3, by 5, by 7…

$100 \div 2$ $50 \div 2$ $25 \div 5$ $5 \div 5$ $100 = 2 \times 2 \times 5 \times 5 = 2^2 \cdot 5^2$

Example 2: Represent your numbers like the product of just prime numbers

$50 = 2 \times 25 = 2 \times 5 \times 5 = 2 \times 5^2$

Grade 8 Chapter 5 Lesson 1

Table 6

Tests for division

Test for division is a way to know, without doing division, if some other number can divide a number evenly or with a remainder of zero.
A number is divisible by another number when there is no remainder.

Test for 2

A number is divisible by 2 if its last digit is even digits (0, 2, 4, 6, 8)
If a number can divide by 2 this number called even. Example: 1,234

Test for 5

A number is divisible by 5 if its last digit is 0 or 5
Example: 100 can be divided by 5 $100 \div 5 = 20$

Test for 10

A number is divisible by 10 if its last digits have one or more zeros
Example: 240 can be divided by 10 $240 \div 10 = 24$

Test for 25

A number is divisible by 25 if the last two digits to the right are 00, 25, 50, 75. Example: 100, 250, 675, 925 $100 \div 25 = 4$

Test for 4

A number is divisible by 4 if the last two digits to the right of the number are 00 or make a new number that can be divided by 4.
Example: 100, 728, 540 $100 \div 4 = 25$ $28 \div 4 = 7$ $40 \div 4 = 10$

Test for 8

A number is divisible by 8 if the last three digits to the right of number are 000 or make a new number that can be divided by 8.
Example: 1000, 2160, 6240 $1000 \div 8 = 125$ $160 \div 4 = 40$ $240 \div 4 = 60$

Test for 3

A number is divisible by 3 if the sum of its digits is divisible by 3

$342 = 300 + 40 + 2 = 3\times100 + 4\times10 + 2 = 3(99+1) + 4(9+1) + 2$

$3 \times 99 + 3 + 4\times9 + 4 + 2 = 3\times99 + 4\times9 + (3+4+2)$

Example: 342 $3+4+2 = 9$ $9 \div 3 = 3$ $342 \div 3 = 114$

Test for 9

A number is divisible by 9 if the sum of its digits is divisible by 9.

$468 = 400 + 60 + 8 = 4\times100 + 6\times10 + 8 = 4(99+1) + 6(9+1) + 8$

$4 \times 99 + 4 + 6\times 9 + 6 + 8 = 4\times99 + 6\times9 + (4 + 6 + 8)$

Example 468 $4+6+8 = 18$ $18 \div 9 = 2$ $468 \div 9 = 52$

Grade 8 Chapter 5 Lesson 1

Table 7

Tests for division

Test for 6

A number is divisible by 6 if it is divisible by both 2 and 3.

Example: 246 Last digit 6 $2+4+6=12$ $12 \div 3 = 4$ $246 \div 6 = 41$

Test for 15

A number is divisible by 15 if it is divisible by both 5 and 3.

Example: 465 Last digit 5 $4 + 6 + 5 = 15$ $15 \div 3 = 5$ $465 \div 15 = 31$

Test for Sum

If each addend can divide by that number the sum divisible by a number

$(a + b) \div c = a \div c + b \div c$ $(18 + 12) \div 3 = 18 \div 3 + 12 \div 3 = 6 + 4 = 10$

Test for Difference

If minuend and subtrahend can divide by a number the difference divisible by a number $(a - b) \div c = a \div c - b \div c$ $(18-12) \div 3 = 18 \div 3 - 12 \div 3 = 6-4 = 2$

Test for Product

If one of factors divide by number the product divisible by a number

$(A \times B) \div C = (A \div C) \times B$ $(6 \times 5) \div 3 = (6 \div 3) \times 5 = 2 \times 5 = 10$

$(A \times B) \div C = A \times (B \div C)$ $(7 \times 6) \div 3 = 7 \times (6 \div 3) = 7 \times 2 = 14$

Mental Math Do operations mentally

If numbers can divide evenly, try to multiply or divide easily

To multiply a number by 5, divide this number by 2, and multiply that result by 10. $16 \times 5 = 16 \div 2 \times 10 = 8 \times 10 = 80$ $24 \times 5 = 24 \div 2 \times 10 = 12 \times 10 = 120$

To multiply a number by 25, divide this number by 4, and multiply that result by 100. $16 \times 25 = 16 \div 4 \times 100 = 4 \times 100 = 400$ $24 \times 25 = 24 \div 4 \times 100 = 6 \times 100 = 600$

To multiply a number by 125, divide this number by 8, and multiply result by 1000. $16 \times 125 = 16 \div 8 \times 1000 = 2 \times 1000 = 2000$

$24 \times 125 = 24 \div 8 \times 1000 = 3 \times 1000 = 3000$

To divide a number by 5, divide this number by 10, and multiply that result by 2. $50 \div 5 = 50 \div 10 \times 2 = 5 \times 2 = 10$ $80 \div 5 = 80 \div 10 \times 2 = 8 \times 2 = 16$

To divide a number by 25, divide this number by 100, and multiply that result by 4. $500 \div 25 = 500 \div 100 \times 4 = 5 \times 4 = 20$ $800 \div 25 = 800 \div 100 \times 4 = 8 \times 4 = 32$

To divide a number by 125, divide this number by 1000, and multiply that result by 8. $5000 \div 125 = 5000 \div 1000 \times 8 = 5 \times 8 = 40$

$8000 \div 125 = 8000 \div 1000 \times 8 = 8 \times 8 = 6$

Grade 8 Chapter 5 Lesson 2

Table 8 Divisors and Multiples

Greatest Common Factor (in short GCF)

The Greatest Common Factor is the greatest number by which all given numbers can be divided.

1) Find the GCF of numbers 12 and 18 (÷ 1, 2, 3, 4, …)
Divisors of number 12 are 1, 2, 3, 4, 6, 12
Divisors of number 18 are 1, 2, 3, 6, 9, 18
Common divisors are 1, 2, 3, 6 Greatest common divisors are 6
GCD (12, 18) = 6 $12 \div 6 = 2$ $18 \div 6 = 3$

2) If greatest of the numbers can be divided evenly by the smallest number, then the greatest common factor of numbers is equal to the smallest of given numbers: Find the GCF of numbers 12 and 4
Factors of 12 are 1, 2, 3, 4, 6, 12 Factors of 4 are 1, 2, 4
Common factors are 1, 2, 4 Greatest common factors is 4
GCF (12, 4) = 4 $12 \div 4 = 3$

3)The greatest common factor of prime numbers equal one:
GCF (5,7) = 1

Least Common Multiples (in short LCM)

The least number is the smallest number which can be evenly divided by each of the given numbers

1) Find the LCM of numbers 12 and 18 (× 1, 2, 3, 4, 5…)
Multiples of 12 are 12, 24, 36, 48, 60, 72…
Multiples of 18 are 18, 36, 54, 72…
Common multiples are 36, 72… Least common multiple is 36
LCM (12, 18) = 36 $36 \div 12 = 3$ $36 \div 18 = 2$

2) If the greatest of given numbers is evenly divided by the smallest number, the LCM of the numbers is equal to the greatest of the given numbers: Find the LCM of numbers 4, 12
Multiples of 4 are 4, 8, 12, 16, 20, 24… Multiples of 12 are 12, 24…
Common multiples are 12, 24… LCM (4, 6, 12) = 12

2) The least common multiple of prime numbers is equal to the
3) product of the numbers. Find the LCM of numbers 2 and 5
Multiples of 2 are 2, 4, 6, 8, 10… Multiples of 5 are 5, 10…

Grade 8 Chapter 5 Lesson 2

Table 9
Decimals
Properties of Decimal

$4 = 4. = 4.0 = 4.00$ $5.00 = 5$

Rounding of Decimals

$2.83 \approx 2.8$ $2.38 \approx 2.4$

Comparison of Decimals

$1.48 < 3.6$ $0.5 > 0.05$

Addition and Subtraction of Decimals

$1.25 + 0.80 = 2.05$ $1.25 - 0.80 = 0.45$

Multiplication of Decimals

$0.24 \times 10 = 2.4$ $0.24 \times 100 = 24$ $0.24 \times 1000 = 240$

$2.5 \times 4 = 10$ $1.25 \times 8 = 10$ $0.5 \times 0.2 = 0.1$

Powers of Decimals

$3.5^2 = 3.5 \times 3.5 = 12.25$ $0.5^3 = 0.5 \times 0.5 \times 0.5 = 0.25 \times 0.5 = 0.125$

Division of Decimals

$5 \div 10 = 0.5$ $5 \div 100 = 0.05$ $5 \div 1000 = 0.005$

$2 \div 5 = 0.4$ $0.2 \div 5 = 0.04$ $0.02 \div 5 = 0.004$

$2 \div 0.5 = 20 \div 5 = 4$ $0.2 \div 0.5 = 2 \div 5 = 0.4$ $0.02 \div 0.5 = 0.2 \div 5 = 0.04$

Order of Operations with the Same Level

$0.8 + 0.6 + 0.2 = 0.8 + 0.2 + 0.6 = 1.6$ $2.5 \times 6 \times 4 = 2.5 \times 4 \times 6 = 60$

$0.8 - 0.2 + 0.4 = 0.6 + 0.4 = 1$ $1.6 \div 4 \times 2 = 0.4 \times 2 = 0.8$

Order of Operations of Two Levels

$0.2 \times 4 - 0.2 = 0.8 - 0.2 = 0.6$ $1.8 \div 9 + 0.4 = 0.2 + 0.4 = 0.6$

$0.6 + 0.4 \times 2 = 0.6 + 0.8 = 1.4$ $0.8 - 1.5 \div 3 = 0.8 - 0.5 = 0.3$

Order of Operations in Expression with Parentheses

$(0.5 + 0.4) \times 2 = 0.9 \times 2 = 1.8$ $3 \times (0.9 - 0.7) = 3 \times 0.2 = 0.6$

$(1.8 - 0.4) \div 7 = 1.4 \div 7 = 0.2$ $2 \div (0.4 - 0.2) = 2 \div 0.2 = 10$

Groping Symbols: 1) Parentheses ( ) 2) Bracket [ ] 3) Brace { }

Order of Five Operations

Grouping Symbols (1) [2] {3} Exponents (Power) a^n

Multiplication or Division from left to right $\times$ or $\div$ L $\rightarrow$ R

Addition or Subtraction from left to right + or – L $\rightarrow$ R

Grade 8 Chapter 5 Lesson 2

Table 10

Order of Four Operations

An arithmetic expression contains numbers, sign of operations and sometimes parentheses. Result of all operations called value of expressions. Expressions with numbers can be in different form:
as a sum $8 + 4$, a difference $8 - 4$, a product 8×4, or a quotient $8 \div 4$.
Operations of the first level are addition (+) and subtraction (-)
Operations of the second level are multiplication (×) and division (÷)
Operations of the third level are power (a^n), root, and logarithm

Order of Operations of the Same Level (+, -) or (×, ÷)

An arithmetic expression without parenthesis, that has just addition or just multiplication, add or multiply in any order:

$80 + 60 + 20 = 80 + 20 + 60 = 160$ $\quad$ $25 \times 6 \times 4 = 25 \times 4 \times 6 = 600$

$11 + 22 + 19 + 28 = (11 + 19) + (22 + 28) = 30 + 50 = 80$

$4 \times 5 \times 25 \times 2 = (4 \times 25) \times (5 \times 2) = 100 \times 10 = 1000$

An expression without parenthesis that has operations with the same level of operations (+, -) or (×, ÷), work in order, from left to the right:

$80 - 20 + 40 = 60 + 40 = 100$ $\quad$ $160 \div 4 \times 2 = 40 \times 2 = 80$

$38 - 8 + 32 - 2 = 30 + 32 - 2 = 62 - 2 = 60$

$64 \div 8 \times 4 \div 2 = 8 \times 4 \div 2 = 32 \div 2 = 16$

Order of Operations of Two Levels (+, -, ×, ÷)

To work an expression without parentheses that has operations of two levels of operations, at first work in order with the second level of operation (×, ÷) from left to right, then work in order with the first level of operations (+, -,) from left to right

$60 + 40 \times 2 = 60 + 80 = 140$ $\quad$ $180 \div 9 + 40 = 20 + 40 = 60$

$20 \times 4 - 20 = 80 - 20 = 60$ $\quad$ $80 - 150 \div 3 = 80 - 50 = 30$

$12 + 18 \times 2 \div 2 = 12 + 36 \div 2 = 12 + 18 = 30$

$18 - 24 \div 6 \times 2 = 18 - 4 \times 2 = 18 - 8 = 10$

$16 \times 4 \div 2 - 6 = 64 \div 2 - 6 = 32 - 6 = 26$

$18 \div 9 \times 2 + 4 = 2 \times 2 + 4 = 4 + 4 = 8$

Grade 8 Chapter 5 Lesson 2
Table 11

Order of Operations in an Expression with Parentheses

In an expression with parentheses: the first step is to work with the operations in parentheses, and the second step is to work before and after parentheses from left to right:

$(50 + 40) \times 2 = 90 \times 2 = 180$

$3 \times (90 - 70) = 3 \times 20 = 60$

$(180 - 40) \div 7 = 140 \div 7 = 20$

$200 \div (40 - 20) = 200 \div 20 = 10$

$15 \div 3 \times (5 + 4) \times 2 = 15 \div 3 \times 9 \times 2 = 5 \times 9 \times 2 = 45 \times 2 = 90$

$20 \div (4 - 2) \times 3 = 20 \div 2 \times 3 = 10 \times 3 = 30$

$6 \times (9 - 7) \div 3 = 6 \times 2 \div 3 = 12 \div 3 = 4$

$(18 - 4) \div 7 \times 5 = 14 \div 7 \times 5 = 2 \times 5 = 10$

Grouping Symbols

Expressions have three kinds of grouping symbols: first do the operations in parentheses, second do the operations - in brackets, and third do the operations - in braces:

Parentheses ( ) **Brackets [ ]** **Braces { }**

Order of Five Operations

Expressions have five operations: first do the operations of the third level (a^n); second, do the operations of the second level (× or ÷), and lastly, do the operations of the first level (+ or -), in order as they occur from left to right.

Remember (in short):

Grouping Symbols	**(1)**	**[2] {3}**
Exponents (Power)		a^n
Multiplication or Division from left to right	**× or ÷**	**L → R**
Addition or Subtract from left to right	**+ or –**	**L → R**

Example 1

$6^2 \div 2^2 \times 3 - 3^2 \times 6 \div 3 = 36 \div 4 \times 3 - 9 \times 6 \div 3 = 9 \times 3 - 54 \div 3 = 27 - 18 = 9$

Example 2

$50 - \{200 \div [60 - (30 + 20)]\} = 50 - \{200 \div [60 - 50]\} = 50 - 200 \div 10 =$
$50 - 20 = 30$

Grade 8 Chapter 5 Lesson 2

Table 12

Fractions

Properties of Fraction

$\frac{a}{b}=\frac{a\cdot n}{b\cdot n}$ $\frac{a\div n}{b\div n}=\frac{a}{b}$

$\frac{2}{3}=\frac{2\times 6}{3\times 6}=\frac{12}{18}$ $\frac{60}{75}=\frac{60\div 3}{75\div 3}=\frac{20}{25}=\frac{20\div 5}{25\div 5}=\frac{4}{5}$

Comparison of Fractions

$\frac{6}{11}>\frac{7}{22}$ $\frac{6}{11}=\frac{6\cdot 2}{11\cdot 2}=\frac{12}{22}$ $\frac{12}{22}>\frac{7}{22}$

Addition of Fractions

$\frac{a}{c}+\frac{b}{c}=\frac{a+b}{c}$ $\frac{2}{5}+\frac{3}{8}+\frac{7}{20}=\frac{2\cdot 8}{5\cdot 8}+\frac{3\cdot 5}{8\cdot 5}+\frac{7\cdot 2}{20\cdot 2}=\frac{16}{40}+\frac{15}{40}+\frac{14}{40}=\frac{45}{40}=\frac{9}{8}=1\frac{1}{8}$

Subtraction of Fractions

$\frac{a}{c}-\frac{b}{c}=\frac{a-b}{c}$ $\frac{3}{10}-\frac{1}{4}=\frac{6}{20}-\frac{5}{20}=\frac{1}{20}$ $\frac{1}{12}-\frac{1}{18}=\frac{3}{36}-\frac{2}{36}=\frac{3-2}{36}=\frac{1}{36}$

Multiplication of Fractions

$\frac{a}{b}\times\frac{c}{d}=\frac{a\times c}{b\times d}=\frac{ac}{bd}$ $\frac{4}{6}\times\frac{9}{8}=\frac{4\times 9}{6\times 8}=\frac{1\times 3}{2\times 2}=\frac{3}{4}$

$\frac{a}{b}\cdot c=\frac{a}{b}\cdot\frac{c}{1}=\frac{a\cdot c}{b}$ $\frac{3}{8}\times 4=\frac{3}{8}\times\frac{4}{1}=\frac{3\times 1}{2\times 1}=\frac{3}{2}=1\frac{1}{2}$

$a\cdot\frac{b}{c}=\frac{a}{1}\cdot\frac{b}{c}=\frac{ab}{c}$ $8\times\frac{5}{4}=\frac{8}{1}\times\frac{5}{8}=\frac{5}{1}=5$

Division of Fractions

$\frac{a}{b}\div\frac{c}{d}=\frac{a}{b}\cdot\frac{d}{c}=\frac{ad}{dc}$ $\frac{5}{6}\div\frac{10}{3}=\frac{5}{6}\times\frac{3}{10}=\frac{1\times 1}{2\times 2}=\frac{1}{4}$

$\frac{a}{b}\div c=\frac{a}{b}\cdot\frac{1}{c}=\frac{a}{bc}$ $\frac{3}{4}\div 6=\frac{3}{4}\div\frac{6}{1}=\frac{3}{4}\times\frac{1}{6}=\frac{1}{8}$

$a\div\frac{b}{c}=\frac{a}{1}\cdot\frac{c}{b}=\frac{ac}{b}$ $6\div\frac{3}{4}=\frac{6}{1}\times\frac{4}{3}=\frac{8}{1}=8$

Power of Fractions

$\left(\frac{a}{b}\right)^n=\frac{a^n}{b^n}$ $\left(\frac{1}{3}\right)^2=\frac{1^2}{3^2}=\frac{1}{9}$

Grade 8 Chapter 5 Lesson 2

Table 13

Mixed Numbers

Mixed Number $2+\frac{1}{2}=2\frac{1}{2}$ $\frac{1}{4}+1=1\frac{1}{4}$ $1\frac{3}{4}=\frac{4\cdot 1+3}{4}=\frac{7}{4}$

Comparison of Mixed Numbers $1\frac{1}{3}<1\frac{5}{6}$ $1\frac{2}{6}<1\frac{5}{6}$

Addition of Mixed Numbers

$1\frac{1}{8}+2\frac{3}{8}=3\frac{4}{8}=3\frac{1}{2}$

$1\frac{1}{3}+2\frac{3}{9}=1\frac{1}{3}+2\frac{1}{3}=3\frac{2}{3}$

$1\frac{2}{6}+2\frac{1}{9}=1\frac{1}{3}+2\frac{1}{9}=1\frac{3}{9}+2\frac{1}{9}=3\frac{4}{9}$

$1\frac{1}{2}+1\frac{1}{3}+2\frac{1}{6}=1\frac{3}{6}+1\frac{2}{6}+2\frac{1}{6}=4\frac{6}{6}=5$

Subtraction of Mixed Numbers

$6\frac{7}{8}-5\frac{3}{8}=1\frac{4}{8}=1\frac{1}{2}$

$4\frac{3}{4}-2\frac{1}{4}=2\frac{2}{4}=2\frac{1}{2}$

$3\frac{1}{2}-2=1\frac{1}{2}$

$1-\frac{2}{3}=\frac{3}{3}-\frac{2}{3}=\frac{1}{3}$

$4-\frac{5}{8}=3\frac{8}{8}-\frac{5}{8}=3\frac{3}{8}$

$8-3\frac{5}{6}=7\frac{6}{6}-3\frac{5}{6}=4\frac{1}{6}$

$6\frac{2}{7}-2\frac{5}{7}=5\frac{9}{7}-2\frac{5}{7}=3\frac{4}{7}$

$5\frac{1}{3}-2\frac{5}{6}=5\frac{2}{6}-2\frac{5}{6}=4\frac{8}{6}-2\frac{5}{6}=2\frac{3}{6}=2\frac{1}{2}$

Multiplication of Mixed Numbers

$3\frac{1}{3}\cdot 2\frac{2}{5}=\frac{10}{3}\cdot\frac{12}{5}=\frac{2\cdot 4}{1\cdot 1}=8$

$\frac{5}{8}\cdot 4\frac{4}{5}\cdot\frac{1}{6}=\frac{5}{8}\cdot\frac{24}{5}\cdot\frac{1}{6}=\frac{3}{6}=\frac{1}{2}$

Division of Mixed Numbers

$1\frac{1}{6}\div 4\frac{2}{3}=\frac{7}{6}\div\frac{14}{3}=\frac{7}{6}\times\frac{3}{14}=\frac{1}{4}$

$3\frac{1}{5}\div 2\frac{2}{5}=\frac{16}{5}\times\frac{5}{12}=\frac{4}{3}=1\frac{1}{3}$

$8\div 4\frac{4}{5}=\frac{8}{1}\div\frac{24}{5}=\frac{8}{1}\times\frac{5}{24}=\frac{5}{3}=1\frac{2}{3}$

$4\frac{1}{3}\div 13=\frac{13}{3}\div\frac{13}{1}=\frac{13}{3}\times\frac{1}{13}=\frac{1}{3}$

Power of Mixed Numbers

$\left(\frac{3}{4}\right)^3=\frac{3^3}{4^3}=\frac{27}{64}$

$\left(1\frac{1}{2}\right)^2=\left(\frac{3}{2}\right)^2=\frac{3^2}{2^2}=\frac{9}{4}=2\frac{1}{4}$

Grade 8 Chapter 5 Lesson 3

Table 14

Fractions and Decimals

Converting Fractions to Terminating Decimals

$10=2\times5 \quad 100=4\times25 \quad 1000 = 8\times125$

$\frac{1}{10} = 0.1 \quad \frac{1}{100} = 0.01 \quad \frac{1}{1000} = 0.001$

$\frac{1}{2} = \frac{1\times5}{2\times5} = \frac{5}{10} = 0.5 \quad \frac{1}{5} = \frac{1\times2}{5\times2} = \frac{2}{10} = 0.2$

$\frac{1}{4} = \frac{1\times25}{4\times25} = \frac{25}{100} = 0.25 \quad \frac{15}{20} = \frac{3}{4} = \frac{3\times25}{4\times25} = \frac{75}{100} = 0.75 \quad 2\frac{1}{8} = 2\frac{1\times125}{8\times125} = 2\frac{125}{1000} = 2.125$

Memorize

$\frac{1}{2} = 0.5 \quad \frac{1}{4} = 0.25 \quad \frac{1}{5} = 0.2 \quad \frac{1}{8} = 0.125 \quad \frac{3}{4} = 0.75$

Converting Terminating Decimals to Mixed Numbers

$1.125 = 1\frac{125}{1000} = 1\frac{125\div5}{1000\div5} = 1\frac{25}{200} = 1\frac{25\div5}{200\div5} = 1\frac{5}{40} = 1\frac{1}{8}$

Combine Operations with Fractions and Decimals

Easy: + and - in decimals, × and ÷ in fractions

$\frac{1}{2} - 0.25 = 0.5 - 0.25 = 0.25 \quad 0.25 \div \frac{1}{8} = \frac{1}{4}\times\frac{8}{1} = 2$

$\frac{2}{3} + 0.6 = \frac{2}{3} + \frac{6}{10} = \frac{2}{3} + \frac{3}{5} = \frac{10}{15} + \frac{9}{15} = \frac{19}{15} = 1\frac{4}{15} \quad 2.8 - 2\frac{2}{3} = 2\frac{8}{10} - 2\frac{2}{3} = 2\frac{4}{5} - 2\frac{2}{3} = \frac{2}{15}$

$\frac{2}{3}\times0.36 = 0.36\times\frac{2}{3} = 0.36\div3\times2 = 0.12\times2 = 0.24 \quad 3.6\div4\frac{1}{2} = 3.6\div4.5 = \frac{3.6}{4.5} = \frac{36}{45} = \frac{4}{5} = 0.8$

Converting Simple Repeating Decimals to Fractions

Yes $\frac{1}{2}, \frac{1}{4}, \frac{1}{5}, \frac{1}{8}, \frac{1}{10}$ No $\frac{1}{3}, \frac{1}{6}, \frac{1}{7}, \frac{1}{9}$ a) $x = 0.3333\ldots\ (\times10) \quad 0.\overline{3} = \frac{3}{9} = \frac{1}{3}$

b) $x = 0.1818\ldots \quad 0.\overline{18} = \frac{18}{99} = \frac{2}{11}$ c) $0.\overline{6} = \frac{6}{9} = \frac{2}{3}$ d) $1.\overline{27} = 1\frac{27}{99} = 1\frac{3}{11}$

Converting Composite Repeating Decimals in the Fractions

a) $x = 0.8333\ldots\ (\times10) \quad 0.8\overline{3} = \frac{83-8}{90} = \frac{75}{90} = \frac{5}{6} = \frac{5}{6}$ b) $0.1\overline{3} = \frac{13-1}{90} = \frac{12}{90} = \frac{2}{15}$

c) $1.1\overline{6} = 1\frac{16-1}{90} = 1\frac{15}{90} = 1\frac{1}{6}$ d) $1.8\overline{3} = 1\frac{83-8}{90} = 1\frac{75}{90} = 1\frac{5}{6}$

Grade 8 Chapter 5 Lesson 3

Table 15

Rational Numbers

Addition and Subtraction of Rational Numbers

$(-1.25) + (-0.8) = -1.25 - 0.8 = -2.05$ $\quad$ $-1.25 + 0.8 = -0.45$

$-1.25 - (-0.8) = -1.25 + 0.8 = -0.45$ $\quad$ $0.8 - 1.25 = -0.45$

Multiplication of Rational Numbers

$(+6)(+2) = +12$ $\quad$ $(-6)(-2) = +12$ $\quad$ $(+)(+) = +$ $\quad$ $(-)(-) = +$

$(+6)(-2) = -12$ $\quad$ $(-6)(+2) = -12$ $\quad$ $(+)(-) = -$ $\quad$ $(-)(+) = -$

$(-0.25)(10) = -2.5$ $\quad$ $(0.25)(-100) = -25$ $\quad$ $(-0.25)(-1000) = 250$

$(-4)(2.5) = -10$ $\quad$ $8(-12.5) = -100$ $\quad$ $(1.25)(-0.8) = -1$

Division of Rational Numbers

$(+6)\div(+2) = +3$ $\quad$ $(-6)\div(-2) = +3$ $\quad$ $(+)\div(+) = +$ $\quad$ $(-)\div(-) = +$

$(-6)\div(+2) = -3$ $\quad$ $(+6)\div(-2) = -3$ $\quad$ $(+)\div(-) = -$ $\quad$ $(-)\div(+) = -$

$(-5)\div 10 = -0.5$ $\quad$ $5 \div (-100) = -0.05$ $\quad$ $(-5)\div(-1000) = 0.005$

$(-2)\div 5 = -0.4$ $\quad$ $0.2 \div (-5) = -0.04$ $\quad$ $(-2)\div(-0.5) = 20\div 5 = 4$

$0.2\div(-0.5) = (-2)\div 5 = -0.4$ $\quad$ $(-1.2)\div 0.8 = (-12)\div 8 = -1.5$

Powers of Rational Numbers

$2^3 = 2\times 2\times 2 = 8$ $\quad$ $(-3)^2 = (-3)(-3) = 9$

$(-2)^3 = -2(-2)(-2) = -8$ $\quad$ $(-2)^4 = (-2)(-2)(-2)(-2) = 16$

$-2^3 = -2\times 2\times 2 = -8$ $\quad$ $-3^2 = -3\times 3 = -9$

$(-0.1)^2 = (-0.1)(-0.1) = 0.01$ $\quad$ $-0.3^2 = -0.09$

Order of Operations

$-25(-6)(-4) = -25\times 4\times 6 = -600$ $\quad$ $-20(-4) + 20 = -80 + 20 = -60$

$-60 - 40(-2) = -60 + 80 = 20$ $\quad$ $-16 \div (-4)(-20) = 4(-20) = -80$

$(-50 - 40)\times 2 = -90\times 2 = -180$ $\quad$ $3(-90 + 70) = 3(-20) = -60$

$-180 \div 9 - 40 = -20 - 40 = -60$ $\quad$ $-80 + 150 \div 3 = -80 + 50 = -30$

$(-180 + 40)\div 7 = -140 \div 7 = -20$ $\quad$ $200 \div (-40 + 20) = 200\div(-20) = -10$

Examples

a) $5(-3^2 - 2^2\times 2^3) = 5(-9 - 4\times 8) = 5(-9 - 32) = 5(-41) = -205$

b) $\{-200 \div [-60 - (-30 - 20)]\} = \{-200 \div [-60 + 50]\} = -200 \div (-10) = 20$

Grade 8 Chapter 5 Lesson 3

Table 16 Laws for operations 1

A formula is a statement of math relations between variables.
Formulas include letters that can represent any numbers.
Equality is two expressions joined by the equal symbol.

Laws of Addition

First Law of Addition (commutative)
Changing the order of two addends does not change the sum:
$A + B = B + A$ $2 + 3 = 5$ $3 + 2 = 5$ $2 + 3 = 3 + 2$
Second Law of Addition (associative)
Changing the grouping of addends does not change the sum:
$A + (B + C) = (A + B) + C$ $8 + (2 + 5) = (8 + 2) + 5 = 10 + 5 = 15$
Use the commutative law of addition as the first step.
Use the associative law of addition as the second step.
Add in any order. Use the two laws together.
$2 + 7 + 3 + 8 = (2 + 8) + (3 + 7) = 10 + 10 = 20$

Laws of Multiplication

First Law of Multiplication (commutative)
Changing the order of two factors does not change the product:
$A\,B = B\,A$ $2 \times 3 = 6$ $3 \times 2 = 6$ $2 \times 3 = 3 \times 2$
Second Law of Multiplication (associative)
Changing the grouping of factors does not change the product:
$A\,(B\,C) = (AB)\,C$ $5 \times (2 \times 4) = (5\times2) \times 4 = 10 \times 4 = 40$
Use the commutative law of multiplication as the first step
Use the associative law of multiplication as the second step
Multiply in any order; use the two laws together:
$5 \times 8 \times 2 = 5 \times 2 \times 8 = 10 \times 8 = 80$

First Distributive Law of Multiplication with Addition

To multiply a number by a sum, multiply that number by each of the addends and add the results
$A\,(B + C) = AB + AC$ $2(3+5) = 2\times 3+ 2\times5= 6 +10=16$
$(A + B)\,C = AC + BC$ $14 \times 5= (10 + 4) \times 5 = 10 \times 5 + 4 \times 5 =50+20=70$

Second Distributive Law of Multiplication with Subtraction

To multiply a number by a difference, multiply that number by the minuend and subtrahend, and subtract the results:
$A\,(B – C) = AB – AC$ $2(10 – 4) = 2 \times 10 – 2 \times 4 = 20 – 8 = 12$
$(A - B)\,C = AC – B\,C$ $(20 - 2) \times 5 = 20 \times 5 – 2 \times 5 = 100 – 10 = 90$
To do mental calculations means to carry out operations by laws or properties.

Grade 8 Chapter 5 Lesson 3

Table 17

Laws for rational numbers 2

Distributive laws of multiplication

1) Whole Numbers (Put A before parentheses)

AB + AC = A (B + C) AB – AC = A (B - C)

If two products have a common factor, put the common factor before parentheses. AB + AC A – common factor

In parentheses, put results of division products by common factor

AB ÷ A = B AC ÷ A = C

$16 \cdot 5 + 4 \cdot 5 = 5\ (16 + 4) = 5\ (20) = 100$

$22 \cdot 5 - 2 \cdot 5 = 5\ (22 - 2) = 5\ (20) = 100$

2) Decimals

a) (B + C) A = AB + AC $(0.4 + 0.1)5 = 0.4 \times 5 + 0.1 \times 5 = 2 + 0.5 = 2.5$

AB + AC = A (B + C) $0.4 \times 5 + 0.1 \times 5 = 5\ (0.4 + 0.1) = 5(0.5) = 2.5$

b) A (B - C) = AB – AC $5\ (0.4 - 0.1) = 5 \times 0.4 - 5 \times 0.1 = 2 - 0.5 = 1.5$

AB – AC = A (B – C) $5 \times 0.4 - 5 \times 0.1 = 5\ (0.4 - 0.1) = 5(0.3) = 1.5$

3) Negative Numbers

If first term negative, put (- A) before parentheses and change each sign inside parentheses: -AB – AC = - A (B + C) - AB + AC = - A (B - C)

a) $-16 \cdot 5 - 4 \cdot 5 = -5\ (16 + 4) = -5\ (20) = -100$

$-22 \cdot 5 + 2 \cdot 5 = -5\ (22 - 2) = -5\ (20) = -100$

b) $-7.5\times5 - 2.5\times5 = -5(7.5 + 2.5) = -5(10) = -50$

$-12.5\ (5) + 2.5(5) = -5(12.5 - 2.5) = -5(10) = -50$

4) To solve problems work by order of operations or change division for multiplication then put A before parentheses.

AB + AC = A (B + C) AB – AC = A (B - C)

a) $2\frac{5}{8} \div \frac{1}{8} + 1\frac{3}{8} \div \frac{1}{8} = 2\frac{5}{8} \times 8 + 1\frac{3}{8} \times 8 = 8(2\frac{5}{8} + 1\frac{3}{8}) = 8(4) = 32$

b) $2\frac{5}{8} \div \frac{1}{8} - 1\frac{3}{8} \div \frac{1}{8} = 2\frac{5}{8} \times 8 - 1\frac{3}{8} \times 8 = 8(2\frac{5}{8} - 1\frac{3}{8}) = 8(1\frac{2}{8}) = 8(\frac{10}{8}) = 10$

5) To solve problems work by order of operations or change division for multiplication then put (- A) before parentheses and change each sign inside parentheses - AB - AC = - A (B + C) - AB + A C = - A (B - C)

a) $-2\frac{5}{8} \div \frac{1}{8} - 1\frac{3}{8} \div \frac{1}{8} = -2\frac{5}{8} \times 8 - 1\frac{3}{8} \times 8 = -8\ (2\frac{5}{8} + 1\frac{3}{8}) = -8\ (4) = -32$

b) $-2\frac{5}{8} \div \frac{1}{8} + 1\frac{3}{8} \div \frac{1}{8} = -2\frac{5}{8} \times 8 + 1\frac{3}{8} \times 8 = -8\ (2\frac{5}{8} - 1\frac{3}{8}) = -8\ (1\frac{2}{8}) = -8\ (\frac{10}{8}) = -10$

Grade 8 Chapter 5 Lesson 3

Table 18

Common Factor for Whole Numbers 1

AB + AC = A (B + C)   AB – AC = A (B - C)

If two products have a common factor, put the common factor before parentheses. AB + AC A–common factor Put A before parentheses

In parentheses, put results of division products by common factor

AB ÷ A = B   AC ÷ A = C

16 · 5 + 4 · 5 = 5 (16 + 4) = 5 (20) = 100

22 · 5 – 2 · 5 = 5 (22 – 2) = 5 (20) = 100

75×5 + 25×5 = 5(75 + 25) = 5(100) = 500

125 (5) - 25(5) = 5(125 - 25) = 5(100) = 500

36 · 5 = (30 + 6) · 5 = 30 · 5 + 6 · 5 = 150 + 30 = 180

8 · 9 + 2 · 9 = 9 · (8 + 2) = 9 · 10 = 90

Memorize Formulas

1)A + (B + C) = (A + B) + C

2)A × (B × C) = (A × B) × C

3)A · (B + C) = A·B + A·C

4)A · (B - C) = A·B - A·C

5)A·B + A·C = A · (B + C)

6)A·B - A·C = A · (B - C)

7)A ÷ C + B ÷ C = (A + B) ÷ C

8)A ÷ C - B ÷ C = (A - B) ÷ C

Examples

a)A (B + C) = AB + AC

5× (40 + 10) = 5 × 40 + 5 × 10 = 200 + 50 = 250

b)A (B - C) = AB – AC

5 × (40 – 10) = 5 × 40 – 5 × 10 = 200 – 50 = 150

c)AB + AC = A (B + C)

40× 5 + 10 × 5 = 5 (40 + 10) = 5(50) = 250

d) AB – AC = A (B – C)

5 × 40 – 5 × 10 = 5 (40 – 10) = 5(30) = 150

Grade 8 Chapter 5 Lesson 3
Table 19
Common Factor for Fractions 2

If first term negative, put (- A) before parentheses and change the sign before each term: -AB – AC = - A (B + C) - AB + AC = - A (B - C)

- 16 · 5 - 4 · 5 = - 5 (16 + 4) = - 5 (20) = - 100

- 22 · 5 + 2 · 5 = - 5 (22 - 2) = - 5 (20) = - 100

Examples: Evaluate

1) $-1\frac{3}{5}(2\frac{1}{8})-1\frac{3}{5}(2\frac{7}{8})=-1\frac{3}{5}(2\frac{1}{8}+2\frac{7}{8})=-\frac{8}{5}(5)=-8$

2) $-\frac{3}{4}(4\frac{2}{3})+\frac{3}{4}(3\frac{1}{3})=-\frac{3}{4}(4\frac{2}{3}-3\frac{1}{3})=-\frac{3}{4}(1\frac{1}{3})=-\frac{3}{4}\times\frac{4}{3}=-1$

3) $\frac{5}{6}(-4\frac{3}{5})+\frac{5}{6}\times1\frac{3}{5}=-\frac{5}{6}(4\frac{3}{5}-1\frac{3}{5})=-\frac{5}{6}(3)=-\frac{5}{2}=-2\frac{1}{2}=-2.5$

4) $-4\frac{1}{2}\times\frac{1}{6}-\frac{1}{6}\times1\frac{1}{2}=-\frac{1}{6}(4\frac{1}{2}+1\frac{1}{2})=-\frac{1}{6}(6)=-1$

5) $-3\frac{3}{4}\times1\frac{1}{6}+2\frac{1}{4}\times1\frac{1}{6}=-1\frac{1}{6}(3\frac{3}{4}-2\frac{1}{4})=-1\frac{1}{6}(1\frac{2}{4})=-\frac{7}{6}\times\frac{3}{2}=-\frac{7}{4}=-1.75$

6) $(-\frac{1}{3}+\frac{1}{5})\div\frac{1}{15}=(-\frac{1}{3}+\frac{1}{5})\times15=-\frac{1}{3}\times\frac{15}{1}+\frac{1}{5}\times\frac{15}{1}=-5+3=-2$

7) $(-1\frac{3}{4}+1\frac{1}{2})\div\frac{1}{8}=(-\frac{7}{4}+\frac{3}{2})\times8=-\frac{7}{4}\times\frac{8}{1}+\frac{3}{2}\times\frac{8}{1}=-14+12=-2$

8) $(-1\frac{1}{2}+1\frac{1}{6})\div\frac{1}{6}=(-\frac{3}{2}+\frac{7}{6})\times6=-\frac{3}{2}\times\frac{6}{1}+\frac{7}{6}\times\frac{6}{1}=-9+7=-2$

9) $(-2\frac{1}{2}-1\frac{1}{3})\div\frac{1}{6}=(-\frac{5}{2}-\frac{4}{3})\times6=-\frac{5}{2}\times\frac{6}{1}-\frac{4}{3}\times\frac{6}{1}=-15-8=-23$

10) $(-1\frac{1}{3}-1\frac{3}{4})\div\frac{1}{12}=(-\frac{4}{3}-\frac{7}{4})\times12=-\frac{4}{3}\times\frac{12}{1}-\frac{7}{4}\times\frac{12}{1}=-16-21=-37$

Grade 8 Chapter 5 Lesson 4

Table 20

Conditional Equations

Six Properties of True Equality: if a = b then

$a + c = b + c$ $a - c = b - c$ $ac = bc$ $a \div c = b \div c$ $c \neq 0$

$a + c = b + c$ $a = b$ $3y + 3 = 6 + 3$ $3y = 6$ $y = 2$

$a - b = c$ $a = c + b$ $3y + 2 = 8$ $3y = 8 - 2$ $3y = 6$ $y=2$

$a + c = b + c$ $a = b$ $3y - 9 = 0$ $3y = 9$ $y = 3$

$a + b = c$ $a = c - b$ $3y + 2 = 8$ $3y = 8 - 2$ $3y = 6$ $y = 2$

Conditional equation has one solution $2y = 8$ $y = 4$

$2x = 0$ $x = 0$ $0x = 2$ $\emptyset$ no solutions

Rule for Solving Equations

1) Open parentheses $a(b + c) = ab + ac$
2) Combine like terms 3) Eliminate number on left side
4) Eliminate letter on right side 5) Multiplication property

Equation with Whole Numbers

$-3(y - 1) = -2(y + 2) + 6$ $- 3y + 3 = -2y - 4 + 6$ $-3y + 3 = -2y + 2$

$-3y + 2y = -3 - 2$ $- y = -5$ $y = 5$

Equations with Decimals

$3(y - 5.8) = 6.5(y + 1.2) - 12.25$ $3y - 17.4 = 6.5y + 7.8 - 12.25$

$3y - 17.4 = 6.5y - 4.45$ $3y = 6.5y + 12.95$

$- 3.5y = 12.95$ $y = - 3.7$

Equations with Fraction (× LCD)

$\frac{1}{6} - \frac{y}{4} = \frac{2-y}{3}$ $\frac{12}{1} \times \frac{1}{6} - \frac{12}{1} \times \frac{y}{4} = \frac{12}{1} \times \frac{2-y}{3}$ $2 - 3y = 4(2 - y)$

$2 - 3y = 8 - 4y$ $-3y + 4y = 8 - 2$ $y = 6$

Proportions

$12 \div 4 = 18 \div 6$ $a \div b = c \div d$ $\frac{12}{4} = \frac{18}{6}$ $3 = 3$ $\frac{a}{b} = \frac{c}{d}$

$12 \times 6 = 72$ $4 \times 18 = 72$ $12 \times 6 = 4 \times 18$ $ad = bc$

$12 \div 18 = 4 \div 6$ $6 \div 4 = 18 \div 12$ $6 \div 18 = 4 \div 12$

$\frac{12}{4} = \frac{18}{y}$ $12y = 4 \cdot 18$ $12y = 72$ $y = 72 \div 12$ $y = 6$

$\frac{12}{y} = \frac{18}{6}$ $18y = 12 \cdot 6$ $18y = 72$ $y = 72 \div 18$ $y = 4$

Grade 8 Chapter 5 Lesson 4

Table 21 Geometry

Right triangles

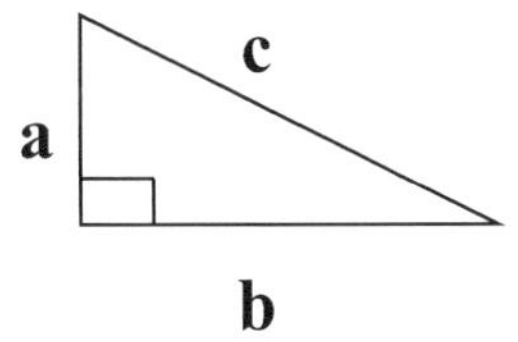

1) Pythagorean Theorem — 1) $c^2 = a^2 + b^2$
2) Area — 2) $A_\Delta = 0.5ab$

Triangles

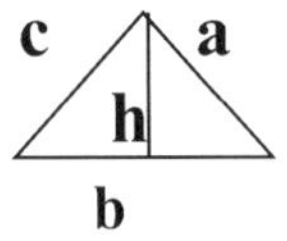

3) Perimeter — 3) $P = a + b + c$
4) Area — 4) $A_\Delta = 0.5bh$

Parallelogram

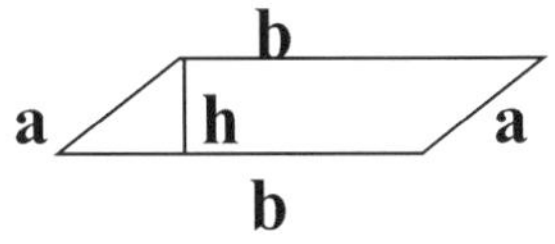

5) Perimeter — 5) $P_{par} = 2a + 2b$
6) Area — 6) $A_{par} = b\ h$

Rhombus

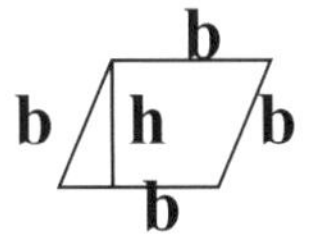

7) Perimeter — 7) $P_{rh} = 4b$
8) Area — 8) $A_{rh} = b \cdot h$

Rectangle

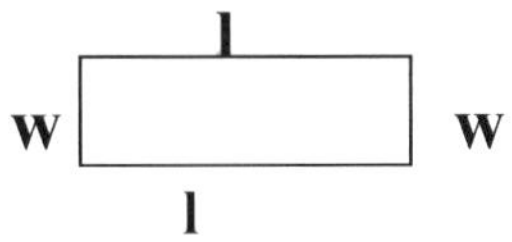

9) Perimeter — 9) $P_{rec} = 2l + 2w$
10) Area — 10) $A_{rec} = l \cdot w$

Square

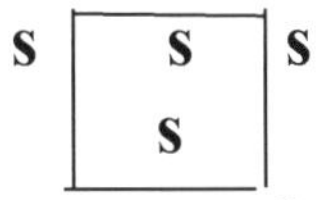

11) Perimeter — 11) $P_{\blacksquare} = 4s$
12) Area — 12) $A_{\blacksquare} = s^2$

Trapezoid

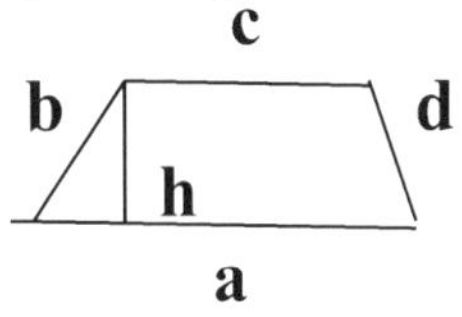

13) Perimeter — 13) $P_{tr} = a + b + c + d$
14) Area — 14) $A_{tr} = 0.5\ h\ (a + c)$

Circle

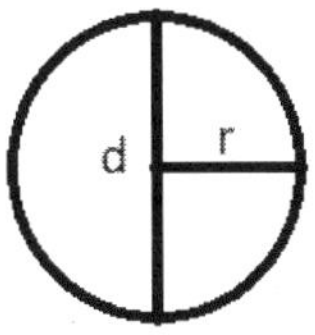

15) Number π in fraction $\approx \frac{22}{7}$ — 16) Number π in decimal ≈ 3.14

17) Diameter $d = 2r$ — 18) Circumference 1 (d) $C = \pi d$

19) Circumference 2 (r) $C = 2\pi r$ — 20) Area $A_c = \pi r^2$

Solid Geometry

Cube 1) $V = s^3$ 2) $SA = 6s^2$ 3) $d = s\sqrt{3}$ Sphere 4) $V = \frac{4}{3}\pi r^3$ 5) $SA = 4\pi r^2$

Rectangular Solid: 6) $V = l \cdot w \cdot h$ 7) $SA = 2lw + 2lh + 2wh$ 8) $d^2 = l^2 + w^2 + h^2$

Regular Prism 9) $V = Base \cdot h$ Regular Pyramid 10) $V = \frac{1}{3} Base \cdot h$

Right Cylinder 11) $V = \pi r^2 h$ Right Cone 13) $V = \frac{1}{3}\pi r^2 h$

Memorize Formulas of Geometry

Grade 8 Chapter 5 Lesson 4

Table 22 Plane Geometry

Right triangles

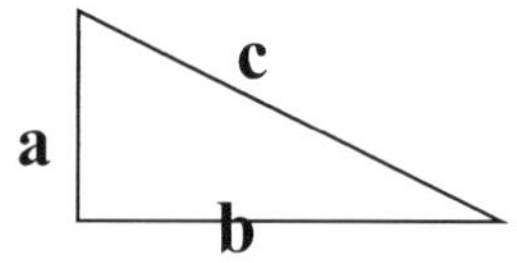

1) Pythagorean Theorem — 1) $c^2 =$

2) Area — 2) $A_\Delta =$

Triangles

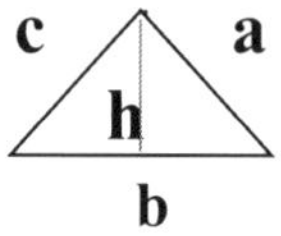

3) Perimeter — 3) $P =$

4) Area — 4) $A_\Delta =$

Parallelogram

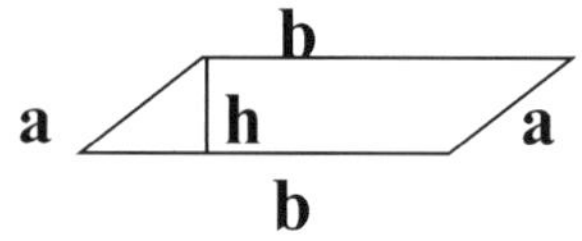

5) Perimeter — 5) $P_{par} =$

6) Area — 6) $A_{par} =$

Rhombus

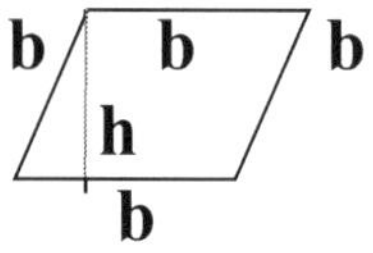

7) Perimeter — 7) $P_{rh} =$

8) Area — 8) $A_{rh} =$

Rectangle

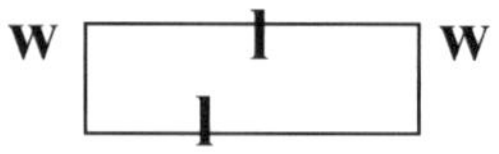

9) Perimeter — 9) $P_{rec} =$

10) Area — 10) $A_{rec} =$

Square

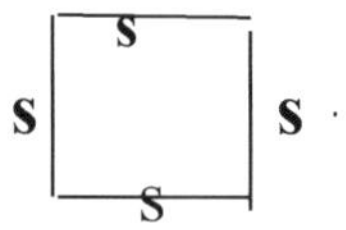

11) Perimeter — 11) $P_\blacksquare =$

12) Area — 12) $A_\blacksquare =$

Trapezoid

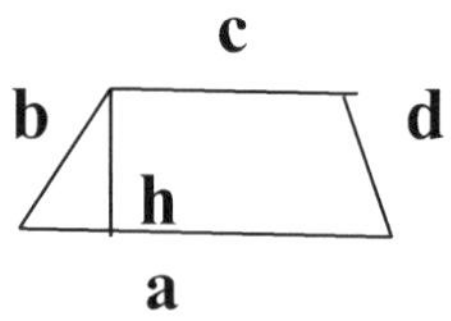

13) Perimeter — 13) $P_{tr} =$

14) Area — 14) $A_{tr} =$

Circle

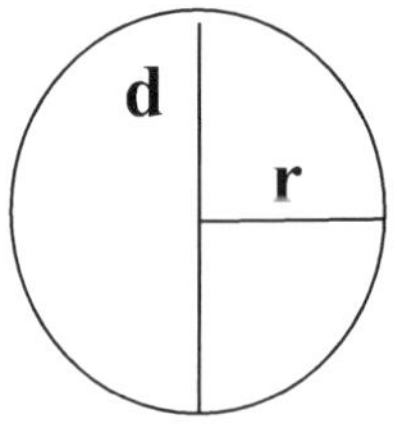

15) Number π in fraction — 15) $\pi \approx$

16) Number π in decimal — 16) $\pi \approx$

17) Diameter — 17) $d =$

18) Circumference 1 (d) — 18) $C =$

19) Circumference 2 (r) — 19) $C =$

20) Area — 20) $A_c =$

Memorize Formulas of Geometry

Grade 8 Chapter 5 Lesson 4

Table 23 Solid Geometry 1

Unit for volume

1 cm = 10 mm — $1\text{ cm}^3 = 1000\text{ mm}^3$

1 yd = 3 ft — $1\text{ yd}^3 = 27\text{ ft}^3$ — s is a side

SA is a surface area — V is a volume — d is a diagonal

Cube

The volume of a cube is equal cube of the its side: $V = s^3$

Surface Area of Cube $SA = 6s^2$

Diagonal of Cube is $B_1D = d$ — $d = s\sqrt{3}$

Example 1

Find the volume, the surface of area, and the diagonal of cube, if s = 4 ft.

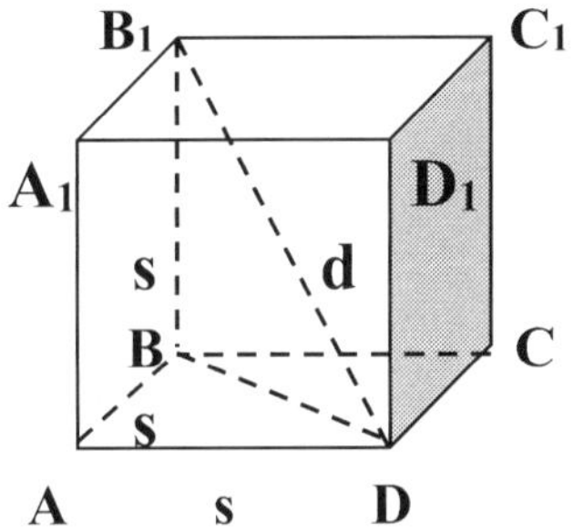

Figure 1

$V = s^3 = 4^3 = 64\text{ ft}^3$

$SA = 6 \cdot 4^2 = 6 \cdot 16 = 96\text{ ft}^2$

$d = s\sqrt{3} = 4\sqrt{3}\text{ ft}$

Rectangular Solid

The volume of a rectangular solid is equal to the product length, width, are height. A rectangular solid has length l, width w, and height h.

Volume of Rectangular Solid $V = l\,w\,h$

Surface Area of Rectangular Solid $SA = 2lw + 2lh + 2wh$

Diagonal of Rectangular Solid $d^2 = l^2 + w^2 + h^2$

Example 2

Find the volume, the surface of area, and the diagonal of rectangular solid, if l = 10 in, w = 6 in, h = 8 in. Figure 2

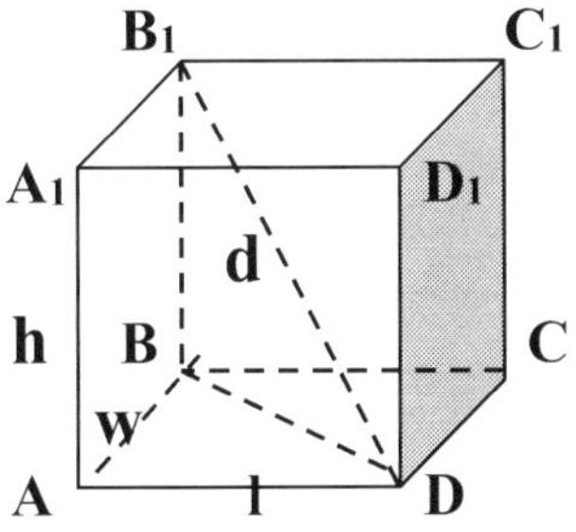

$V = l\,w\,h = 10 \cdot 6 \cdot 8 = 60 \cdot 8 = 480\text{ in}^3$

$SA = 2lw + 2lh + 2wh = 2 \cdot 10 \cdot 6 + 2 \cdot 10 \cdot 8 + 2 \cdot 6 \cdot 8 =$

$20 \cdot 6 + 20 \cdot 8 + 12 \cdot 8 = 120 + 160 + 96 = 376\text{ in}^2$

$d^2 = l^2 + w^2 + h^2 = 10^2 + 6^2 + 8^2 = 100 + 36 + 64 = 100 + 100 = 200$

$d = \sqrt{200} = \sqrt{100 \cdot 2} = 10\sqrt{2}\text{ in}$

Grade 8 Chapter 5 Lesson 4

Table 24 Solid Geometry 2

Sphere

The volume of a sphere is equal to the product four over three of the number π by cube of radius. An example of a sphere is any balls.

Volume of sphere is $V = \frac{4}{3}\pi r^3$ where $\pi \approx \frac{22}{7}$ or $\pi \approx 3.14$

Surface Area of Sphere $SA = 4\pi r^2$ $OB = r$

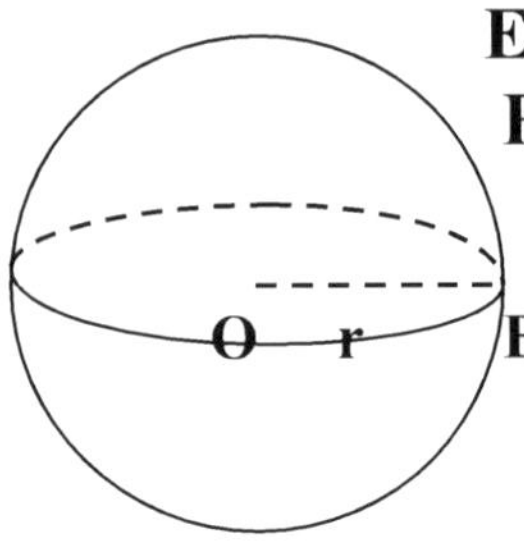

Example 3 Figure 3

Find exact volume and exact surface area of a sphere if the radius is 3 m.

$$V = \frac{4}{3}\pi r^3 = \frac{4}{3}\pi \times 3^3 = \frac{4}{3} \times \frac{27}{1}\pi = 4 \times 9\pi = 36\pi \ m^3$$

$$SA = 4\pi r^2 = 4\pi \times 3^2 = 4 \times 9\pi = 36\pi \ m^2$$

Regular Prism

The volume of a regular prism is equal to the product of the base and height:

$V = Base \cdot h$ Base means area of triangles ABC or $A_1B_1C_1$

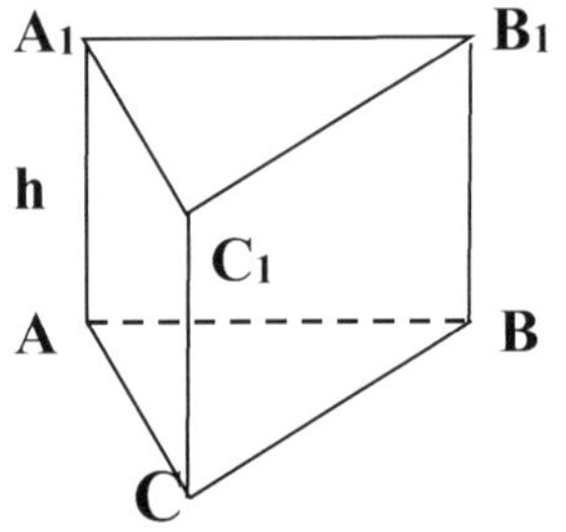

Figure 4

Example 4 Find the volume of a regular prism if the base is 25 m^2 and the height is 10 m. Base=Area of triangle ABC = 25 m^2 h = 10 m

$V = B_{base} \cdot h = 25 \cdot 10 = 250 \ m^3$

Right Circular Cylinder

The volume of a right circular cylinder is equal to the product of the number π, and the square of radius by height: $V = \pi r^2 h$

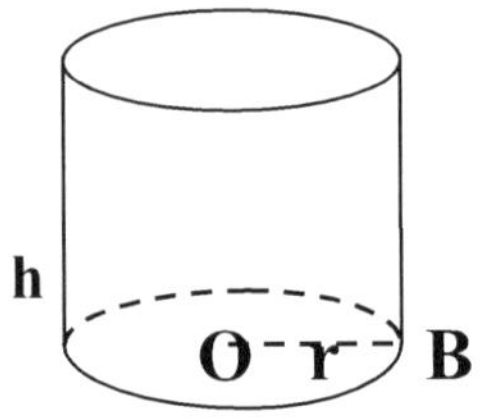

Figure 5

Example 5 Find volume of a right circular cylinder if r = 7 in h = 10 in and $\pi \approx \frac{22}{7}$

$$V = \pi r^2 h = \frac{22}{7} \cdot 7^2 \cdot 10 = \frac{22}{7} \cdot \frac{49}{1} \cdot 10 = 22 \cdot 7 \cdot 10 = 1540 \ in^3$$

Grade 8 Chapter 5 Lesson 4

Table 25

Solid Geometry 3

Regular Pyramid

The volume of a regular pyramid is equal to the product of one-third of base by height: $V = \frac{1}{3}\text{Base} \cdot h$ **Base means area of triangles ABC**

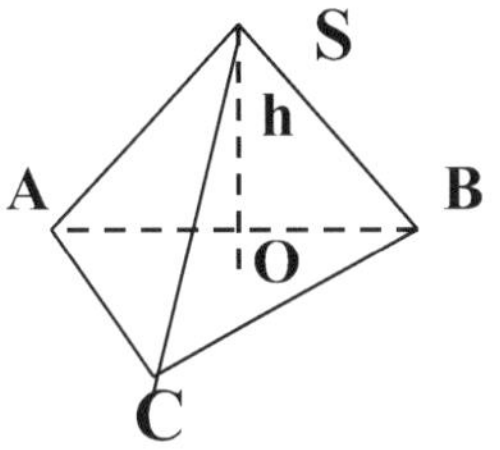

Figure 6

Example 6 **Find the volume of a regular pyramid if the base is 120 m^2 and height is 5 m. The base is the area of the triangle ABC = 120 m^2**

$h = 5$ m $\quad V = \frac{1}{3}\text{Base} \cdot h = \frac{1}{3} \cdot 120 \cdot 5 = 40 \cdot 5 = 200 \text{ m}^3$

Right Circular Cone

The volume of a right circular cone is equal to the product of one-third of square radius by height: $V = \frac{1}{3}\pi r^2 h \quad \pi \approx 3.14 \quad$ or $\pi \approx \frac{22}{7}$

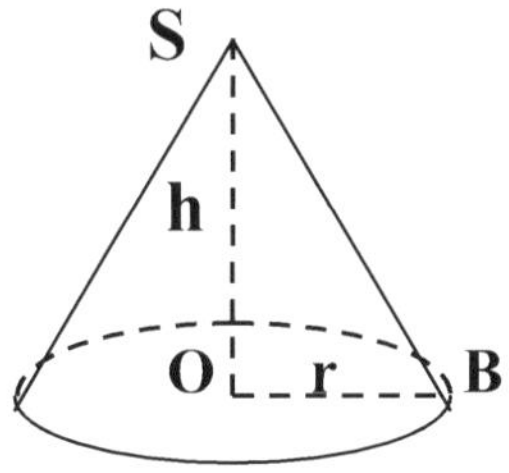

Figure 7

Example 7 **Find the volume of a right circular cone if its height is 10 ft and radius is 3 ft:** $h = 10$ ft $\quad r = 3$ ft

$V = \frac{1}{3}\pi r^2 h = \frac{1}{3}\pi \cdot 3^2 \cdot 10 = \frac{1}{3} \cdot 9 \cdot 10\pi = 3 \cdot 10\pi = 30\pi = 30\ (3.14) = 94.2 \text{ ft}^3$

Memorize Formulas of Geometry

Cube 1) V = 2) SA = 3) d = **Sphere** 4) V = 5) SA =

Rectangular Solid: 6) V = 7) SA = 8) d^2 =

Regular Prism 9) V = **Regular Pyramid** 10) V =

Right Cylinder 11) V = **Right Cone** 12) V =

Grade 8 Chapter 5 Lesson 5

Table 26

Word Problems

1) Decimals

a) To find the amount by known base, multiply the base by a decimal:

Find the amount 0.6 of 40 $40 \times 0.6 = 24$

b) To find the base by known amount, divide the amount by a decimal:

Finding the base: 0.6 is equivalent \$ 24 $24 \div 0.6 = 40$

c) To find the ratio between two numbers, divide the numbers.

Find the ratio 24 from 40 $24 \div 40 = 0.6$

2) Fractions

a) To find the amount by known base, multiply the base by a fraction.

Finding the amount: $\frac{3}{5}$ of 40 $40 \times \frac{3}{5} = \$24$

b) To find the base by known amount, divide amount by a fraction.

Finding the base: $\frac{3}{5}$ equivalent 24 $24 \div \frac{3}{5} = \$40$

c) To find the ratio between two numbers, divide the numbers.

Find the ratio: 24 from \$40 $24 \div 40 = \frac{24}{40} = \frac{3}{5}$

3) Uniform Motion: Distance - d Time - t Rate – r (Speed)

a) To find distance, multiply rate by time: $d = rt$

What distance did a man go for 2 hours if his rate was 5 mi/hr.?

$d = 5 \cdot 2 = 10$ mi

b) To find time, divide distance by speed. $t = d \div r$

How much time does a man needs if he runs 8 mi with a rate 4 mi/hr.?

$t = 8 \div 4 = 2$ hrs.

c) To find speed, divide distance by time. $r = d \div t$

What is the rate a girl has, if she runs 6 mi for 2 hours?

$r = 6 \div 2 = 3$ mi / hr.

4) Directly proportional quantities

If 4 apples sell for \$1.32. How much do 11 apples cost?

$132 \div 4 = 33$ c $33 \times 11 = 363$ c = \$3.63

Grade 8 Chapter 5 Lesson 5

Table 27 Percent Problems

Solve Percent Problems by Equations ("is" and "of")

1) Find A: What numbers is 25% of 40? $A = 0.25 \cdot 40$ $\underline{A = 10}$

2) Find B: 10 is 25% of what number?

$10 = \frac{25}{100}B$ $\quad \frac{10}{0.25} = \frac{0.25B}{0.25}$ $\quad \frac{1000}{25} = B$ $\quad \underline{B = 40}$

3) Find R: What percent of 40 is 10?

$\frac{R}{100} \times 40 = 10$ $\quad 100 \times \frac{R}{100} \times 40 = 10 \times 100$ $\quad 40R = 1000$ $\quad \underline{R = 25\%}$

Solve Percent Problems by Proportion

1) Find A: Michael solves 80% of 20 problems in test.
How many problems he solved?

$\frac{A}{B} = \frac{R}{100}$ $\quad \frac{A}{20} = \frac{80}{100}$ $\quad \frac{A}{20} = \frac{8}{10}$ $\quad 10A = 160$ $\quad A = \underline{16 \text{ problems}}$

2) Find B: David solved 16 problems in test, it is equal 80% of all test.
How many problems were in test?

$\frac{A}{B} = \frac{R}{100}$ $\quad \frac{16}{B} = \frac{80}{100}$ $\quad \frac{16}{B} = \frac{8}{10}$ $\quad 8B = 160$ $\quad \underline{B = 20 \text{ problems}}$

3) Find R: What percent of 20 is 16?

$\frac{A}{B} = \frac{R}{100}$ $\quad \frac{16}{20} = \frac{R}{100}$ $\quad \frac{4}{5} = \frac{R}{100}$ $\quad 5R = 400$ $\quad \underline{R = 80\%}$

Solve Percent Problems by Formulas

1) Find A: Olivia had \$ 30, she spent 70% of all her money.
How much money she spent? To find the amount, multiply the base by the percent, which converted to decimal.

Formula 1 $A = B \cdot \frac{R}{100}$ $\quad A = 30 \cdot \frac{70}{100}$ $\quad A = 30 \cdot 0.7$ $\quad A = \underline{\$ 21}$

2) Find B: Anna spent \$ 21, it is 70% of all her money.
How much money she had? To find the base, divide the amount by the percent, which converted to decimal.

Formula 2 $B = A \div \frac{R}{100}$ $\quad B = 21 \div \frac{70}{100}$ $\quad B = 21 \div 0.7 = 210 \div 7 = \underline{\$ 30}$

3) Find R: Vicki had \$ 30, she spent \$ 21. What percent of money she spent?
To find a percent, divide the numbers and multiply the result by 100%.

Formula 3 $R = \frac{A}{B} \cdot 100\%$ $\quad R = \frac{21}{30} \cdot 100\%$ $\quad R = \frac{7}{10} \times 100\%$ $\quad R = \underline{70\%}$

To solve the percent problems, choose one of three ways which is easy for you.

Grade 8 Chapter 5 Lesson 5

Table 28

Uniform Motion 1

The motion of a moving object whose speed and direction do not change is called uniform motion.

Word problem at moving content three quantities

Distance - d Time - t Rate – r (Speed)

Rate – Distance passed for time unit (for 1 hour, 1 minute)

Movement units

Distance (d) measure mi, m, km

Time (t) measure in min, hrs.

Rate (r) measure in mi/hrs., m/min, km/hrs.

Distance d

To find distance, multiply rate by time: d = rt

What distance did a man runs in 2 hours, if his rate was 5 mi/hr.?

$d = 5 \times 2 = 10$ mi

If men drive at the speed of 70mph for 2 hours, how far will he travel?

$d = 70 \times 2 = 140$ mi

Time t

To find time, divide distance by speed: $t = d \div r$

How much time needed if a man went 8 mi with a rate of 4 mi/hrs.?

$t = 8 \div 4 = 2$ hrs.

If a man drives at a speed of 52 mph for 91 mi, how long will he travel?

$t = d \div r = 91$ mi $\div$ 52 mph $= 7 \div 4 = 1.75$ hours

Rate (Speed) r

To find the speed, divide distance by time: $r = d \div t$

How much rate does a girl have, if she runs 6 mi in 2 hours.

$r = 6 \div 2 = 3$ mi / hr.

Woman drove a distance of 252 mi in $4\frac{2}{3}$ hours. What was her speed?

$r = d \div t = 252 \div 4\frac{2}{3} = \frac{252}{1} \times \frac{3}{14} = 18 \times 3 = 54$ mph

Grade 8 Chapter 5 Lesson 5

Table 29

Uniform Motion 2

Find the Unknowns of Uniform Motion

#		Distance	Time	Speed	Answers
1	Pedestrian	X mi	4 hrs.	5 mi/hr.	20 mi
2	Tractor	X mi	2 hrs.	30 mi/hr.	60 mi
3	Train	X mi	3 hrs.	120 mi/hr.	360 mi
4	Plane	X mi	2 hrs.	1000 mi/hr.	2000 mi
5	Skier	26 mi	X hrs.	13 mi/hr.	2 hrs.
6	Train	240 mi	X hrs.	60 mi/hr.	4 hrs.
7	Car	160 mi	X hrs.	80 mi/hr.	2 hrs.
8	Plane	5000 mi	X hrs.	2500 mi/hr.	2 hrs.
9	Bicycle	28 mi	2 hrs.	X mi/hr.	14 mi/hr.
10	Car	210 mi	3 hrs.	X mi/hr.	70 mi/hr.
11	Plane	15000 mi	3 hrs.	X mi/hr.	5000 mi/hr.
12	Eagle	X mi	1 hr.	120 mi/hr.	120 mi
13	Starling	X mi	2 hrs.	80 mi/hr.	160 mi
14	Swallow	X mi	3 hrs.	70 mi/hr.	210 mi
15	Deer	X mi	3 hrs.	10 mi/hr.	30 mi
16	Shark	72 mi	X hrs.	36 mi/hr.	2 hrs.
17	Ship	280 mi	4 hrs.	X mi/hr.	70 mi/hr.
18	Sparrow	120 mi	2 hrs.	X mi/hr.	60 mi/hr.
19	Zebra	45 mi	X hrs.	15 mi/hr.	3 hrs.
20	Boat	X mi	4 hrs.	24 mi/hr.	96 mi
21	Car	The same	2 hrs.	60 mi/hr.	120 mi
22	Car	The same	3 hrs.	X mi/hr.	40 mi/hr.
23		X mi	2 hrs.	60 mi/hr.	120 mi
24		120 mi	X hrs.	60 mi/hr.	2 hrs.
25		120 mi	2 hrs.	X mi/hr.	60 mi/hr.
26		X mi	2 hrs.	40 mi/hr.	80 mi
27		24 mi	X hrs.	6 mi/hr.	4 hrs.
28		300 mi	5 hrs.	X mi/hr.	60 mi/hr.
29		70 m	7 secs.	The same	10 m/sec.
30		90 m	X secs.	The same	9 secs.

Grade 8 Chapter 5 Lesson 5

Table 30

Application of Percent 1

1. **Sales Tax Formula**
   Amount of sales tax = Cost of merchandise × Tax rate
   Example: A graphing calculator cost \$ 100. The sales tax rate is 5.5%.
   Compute the total cost. 5.5% = 0.55 100×0.055 =\$5.5 100 + 5.5 = \$105.5
2. **Percent of commission**
   Amount of commission = Dollar sales × Percent of commission
   Example: Diana earned 5% on a house she sold for \$ 100,000
   Find her dollar commission. 100000× 0.05 = \$ 5000
3. **Percent of decrease**
   Amount of decrease = original value - new value
   Amount of decrease = original value × percent of decrease
   Percent of decrease = amount of decrease ÷ original value ×100%
   Example: A paper shredder was decreased price from \$ 80 to \$ 60
   What is the percent decreased in price?
   80 – 60 = 20 20 ÷ 80 × 100% = 0.25 × 100% = 25%
4. **Percent of discount**
   Amount of decrease = original value - new value
   Amount of decrease = original value × percent of decrease
   Percent of decrease = amount of decrease ÷ original value ×100%
   Example: The Company put suits that usually sell for \$ 250 on sale for \$ 170.
   What the percent discount?
   250 – 170 = \$ 80 80 ÷ 250 × 100% = 0.32 × 100% = 32%
5. **Percent of increase**
   Amount of increase = new value – original value
   Amount of increase = original value × percent of increase
   Percent of increase = amount of increase ÷ original value × 100%
   Example: The sales in one-year Honda Civic increased from 400,000 to 408,000. Compute the percent increase.
   408000 – 400000=8000 8000÷400000 ×100 = 0.2×100=20%
6. **Percent markup**
   Amount of markup = new value – original value
   Amount of markup = original value × percent of markup (rate)
   Percent of markup = amount of markup ÷ original value × 100%
   Example: A college bookstore marks up of math book is 30%.
   What is mark up for math book that has a price of \$ 70.
   \$70 × 0.3 = \$ 21 70 + 21 = \$ 91

Grade 8 Chapter 5 Lesson 5

Table 31

Application of Percent 2

7. Percent of increase

Amount of increase = new value – original value
Amount of increase = original value × percent of increase
Percent of increase = amount of increase ÷ original value × 100%
Example: The sales in one-year Honda Civic increased from 400,000 to 408,000. Compute the percent increase.
$408000 - 400000 = 8000$ $8000 \div 400000 \times 100 = 0.2 \times 100 = 20\%$

8. Percent markup

Amount of markup = new value – original value
Amount of markup = original value × percent of markup (rate)
Percent of markup = amount of markup ÷ original value × 100%
Example: A college bookstore marks up of math book is 30%.
What is mark up for math book that has a price of \$ 70.
$\$70 \times 0.3 = \$\,21$ $70 + 21 = \$\,91$

9. Percent of Profit

Amount profit = Total Amount income - Amount Expenses
Amount profit = Total Amount income × Percent Profit
Percent of profit = Amount expenses ÷ Total Amount income ×100%
Example: The company had expenses of 148,000 and a profit of \$ 12,000. Find the percent profit.
Total amount income = \$ 148,000 + \$ 12, 000= \$ 160, 000
$\frac{12000}{160000} = \frac{n}{100}$ $\frac{12}{160} = \frac{n}{100}$ $160n=1200$ $16n=120$ $4n=30$ $n=7.5\%$

10. Amount of expenses = total amount income × percent of expenses

Example: The company had a total income of \$ 150,000 and 7,500profit last month.
Amount expenses \$ 150000 – 7,500 = 142,500
$\frac{142500}{150000} = \frac{n}{100}$ $\frac{1425}{1500} = \frac{n}{100}$ $1500n=142500$ $15n=1425$ $n = 95\%$

Grade 8 Chapter 5 Lesson 6
Mental Math 1
Add

1) 9 + 1
2) 8 + 2
3) 7 + 3
4) 6 + 4
5) 5 + 5
6) 9 + 2
7) 8 + 3
8) 7 + 4
9) 6 + 5
10) 9 + 3
11) 8 + 4
12) 7 + 5
13) 6 + 6
14) 9 + 4
15) 8 + 5
16) 7 + 6
17) 9 + 5
18) 8 + 6
19) 7 + 7
20) 9 + 6
21) 8 + 7
22) 9 + 7
23) 8 + 8
24) 9 + 8
25) 9 + 9

Grade 8 Chapter 5 Lesson 6
Mental Math 2
Subtract

1) 10 – 1
2) 10 – 2
3) 10 – 3
4) 10 – 4
5) 10 – 5
6) 11 - 2
7) 11 – 3
8) 11 – 4
9) 11 – 5
10) 12 – 3
11) 12 – 4
12) 12 – 5
13) 12 – 6
14) 13 – 4
15) 13 – 5
16) 13 – 6
17) 14 – 5
18) 14 – 6
19) 14 – 7
20) 15 – 6
21) 15 – 7
22) 16 – 7
23) 16 – 8
24) 17 – 8
25) 18 - 9

Grade 8 Chapter 5 Lesson 6
Mental Math 3
Multiply

1) 5×6
2) 6×6
3) 7×6
4) 8×6
5) 9×6
6) 5×7
7) 6×7
8) 7×7
9) 8×7
10) 9×7
11) 5×8
12) 6×8
13) 7×8
14) 8×8
15) 9×8
16) 5×9
17) 6×9
18) 7×9
19) 8×9
20) 9×9
21) 5×10
22) 6×10
23) 7×10
24) 8×10
25) 9×10

Grade 8 Chapter 5 Lesson 6
Mental Math 4
Divide

1) $30 \div 6$
2) $36 \div 6$
3) $42 \div 6$
4) $48 \div 6$
5) $54 \div 6$
6) $35 \div 7$
7) $42 \div 7$
8) $49 \div 7$
9) $56 \div 7$
10) $63 \div 7$
11) $40 \div 8$
12) $48 \div 8$
13) $56 \div 8$
14) $64 \div 8$
15) $72 \div 8$
16) $45 \div 9$
17) $54 \div 9$
18) $63 \div 9$
19) $72 \div 9$
20) $81 \div 9$
21) $50 \div 10$
22) $60 \div 10$
23) $70 \div 10$
24) $80 \div 10$
25) $90 \div 10$

Grade 8 Chapter 5 Lesson 6
Mental Math 5 Exponents 1

1)	11^2	11)	21^2
2)	12^2	12)	22^2
3)	13^2	13)	23^2
4)	14^2	14)	24^2
5)	15^2	15)	25^2
6)	16^2	16)	26^2
7)	17^2	17)	27^2
8)	18^2	18)	28^2
9)	19^2	19)	29^2
10)	20^2	20)	30^2

Grade 8 Chapter 5 Lesson 7
Mental Math 6
Laws for Whole Numbers 3

1) $8\times3 + 2\times3$
2) $7\times5 - 2\times5$
3) $7\times4 + 3\times4$
4) $6\times2 - 4\times2$
5) $4\times5 + 6\times5$
6) $9\times4 - 8\times4$
7) $5\times2 + 5\times8$
8) $3\times7 - 3\times6$
9) $3\times4 + 3\times6$
10) $5\times9 - 5\times8$

11) $8\times75 + 8\times25$
12) $25\times6 - 15\times6$
13) $2\times17 + 2\times13$
14) $75\times4 - 35\times4$
15) $3\times27 + 3\times23$
16) $93\times 2 - 2\times53$
17) $14\times4 + 4\times16$
18) $6\times28 - 6\times18$
19) $29\times3 + 21\times3$
20) $67\times2 - 2\times47$

Grade 8 Chapter 5 Lesson 7
Mental Math 7 Exponents 2

1) 1^3
2) 2^3

3) 3^3
4) 4^3

5) 5^3
6) 6^3

7) 7^3
8) 8^3

9) 9^3
10) 10^3

11) 2^1
12) 2^2

13) 2^3
14) 2^4

15) 2^5
16) 2^6

17) 2^7
18) 2^8

19) 2^9
20) 2^{10}

Grade 8 Chapter 5 Lesson 7
Mental Math 8
Find Prime Factorization

1) 4
2) 9

3) 16
4) 25

5) 36
6) 49

7) 64
8) 81

9) 6
10) 8

11) 10
12) 12

13) 14
14) 15

15) 18
16) 20

17) 24
18) 27

19) 30
20) 32

Grade 8 Chapter 5 Lesson 7
Mental Math 9
Decimals

1) $0.2 + 0$
2) $0 + 1.5$
3) $0.2 + 0.5$
4) $0.5 + 0.2$
5) $1.8 + 0.2$
6) $0.6 - 0$
7) $1.7 - 1.7$
8) $0.5 - 0.2$
9) 0.6×0
10) 0×1.7
11) 1×1.9
12) 0.8×1
13) 0.2×5
14) 0.4×5
15) 5×0.6
16) 0.5×8
17) 1.2×5
18) 5×1.4
19) $0 \div 0.6$
20) $0.7 \div 0$
21) $0.8 \div 1$
22) $1.9 \div 1.9$
23) 0.2^2
24) 0.3^2
25) 0.4^2
26) 0.5^2
27) 0.6^2
28) 0.7^2
29) 0.8^2
30) 0.9^2

Grade 8 Chapter 5 Lesson 7
Mental Math 10
Laws for Decimals 4

1) $8\times7.5+8\times2.5$

2) $3.7\times4+4\times1.3$

3) $2.5\times4-1.5\times4$

4) $5.4\times5-3.4\times5$

5) $7\times3.7+6.3\times7$

6) $4.9\times2+2\times5.1$

7) $7.5\times6-3.5\times6$

8) $3.9\times8-8\times1.9$

9) $3\times2.7+3\times2.3$

10) $2.4\times3+1.6\times3$

11) $9.3 \times 2 - 2 \times 5.3$

12) $5.8 \times 3 + 4.2 \times 3$

13) $4.3 \times 4 + 4 \times 1.7$

14) $9.6 \times 2 - 2.6 \times 2$

15) $5 \times 2.8 - 5 \times 1.8$

16) $3.8 \times 4 - 1.8 \times 4$

17) $2.9 \times 3 + 3.1 \times 3$

18) $0.3 \times 6 + 0.7 \times 6$

19) $6.7 \times 4 - 4 \times 4.7$

20) $8 \times 5.7 - 8 \times 4.7$

Grade 8 Chapter 5 Lesson 7
Mental Math 11
Fractions

1) $\frac{1}{2} \times \frac{2}{5}$

2) $\frac{1}{3} \times \frac{3}{4}$

3) $\frac{5}{8} \times 8$

4) $4 \times \frac{1}{4}$

5) $\frac{1}{2} \div \frac{1}{2}$

6) $\frac{1}{3} \div \frac{1}{6}$

7) $\frac{1}{4} \div \frac{1}{12}$

8) $2 \div \frac{1}{2}$

9) $(\frac{1}{2})^2$

10) $(\frac{1}{5})^2$

11) $(\frac{2}{5})^2$

12) $(\frac{3}{5})^2$

13) $\frac{1}{4} + \frac{3}{4}$

14) $\frac{1}{2} + \frac{3}{2}$

15) $\frac{7}{8} - \frac{5}{8}$

16) $\frac{7}{8} - \frac{3}{8}$

17) $\frac{1}{2} + \frac{1}{4}$

18) $\frac{3}{4} + \frac{1}{2}$

19) $\frac{1}{2} - \frac{1}{4}$

20) $\frac{3}{4} - \frac{1}{2}$

Grade 8 Chapter 5 Lesson 8
Mental Math 12
Mixed numbers

1) $1+\frac{1}{2}$

2) $\frac{1}{4}+2$

3) $1-\frac{1}{2}$

4) $2\frac{1}{2}-1$

5) $2\frac{3}{8}+1\frac{1}{8}$

6) $3\frac{3}{4}+2\frac{1}{4}$

7) $4\frac{3}{4}-2\frac{1}{4}$

8) $3\frac{5}{8}-2\frac{1}{8}$

9) $3\frac{1}{2}+2\frac{1}{2}$

10) $4\frac{1}{2}+2\frac{1}{2}$

11) $2\frac{3}{4}-1\frac{1}{4}$

12) $4\frac{5}{6}-1\frac{2}{6}$

13) $2\frac{3}{8}-1\frac{1}{8}$

14) $3\frac{3}{8}-2\frac{1}{8}$

15) $4\frac{7}{8}-3\frac{3}{8}$

16) $5\frac{5}{8}-4\frac{1}{8}$

17) $2-1\frac{1}{2}$

18) $2-1\frac{3}{4}$

19) $3-1\frac{1}{4}$

20) $3-1\frac{1}{5}$

Grade 8 Chapter 5 Lesson 8
Mental Math 13
Laws for Fractions 5

1) $\frac{1}{2}\times 8+\frac{1}{2}\times 2$

2) $\frac{4}{3}\times 6-\frac{1}{3}\times 6$

3) $\frac{1}{4}\times 8+\frac{3}{4}\times 8$

4) $\frac{6}{5}\times 5-\frac{1}{5}\times 5$

5) $\frac{1}{3}\times 9+\frac{2}{3}\times 9$

6) $\frac{3}{2}\times 8-\frac{1}{2}\times 8$

7) $\frac{1}{3}\times 6+\frac{2}{3}\times 6$

8) $\frac{5}{4}\times 4-\frac{1}{4}\times 4$

9) $\frac{1}{5}\times 5+\frac{4}{5}\times 5$

10) $\frac{7}{6}\times 6-\frac{1}{6}\times 6$

11) $\frac{1}{6}\times 6+\frac{5}{6}\times 6$

12) $\frac{5}{2}\times 4-\frac{1}{2}\times 4$

13) $\frac{3}{7}\times 7+\frac{4}{7}\times 7$

14) $\frac{7}{3}\times 3-\frac{1}{3}\times 3$

15) $\frac{3}{8}\times 8+\frac{5}{8}\times 8$

16) $\frac{5}{4}\times 8-\frac{1}{4}\times 8$

17) $\frac{4}{9}\times 9+\frac{5}{9}\times 9$

18) $\frac{7}{5}\times 5-\frac{2}{5}\times 5$

19) $\frac{3}{5}\times 5+\frac{2}{5}\times 5$

20) $\frac{7}{6}\times 6-\frac{1}{6}\times 6$

Grade 8 Chapter 5 Lesson 8
Mental Math 14
Write the Fractions as Decimals

1) $\frac{3}{10}$

2) $\frac{9}{100}$

3) $\frac{7}{1000}$

4) $\frac{1}{2}$

5) $\frac{12}{24}$

6) $\frac{1}{5}$

7) $\frac{2}{5}$

8) $\frac{3}{5}$

9) $\frac{4}{5}$

10) $\frac{12}{15}$

11) $\frac{1}{4}$

12) $\frac{2}{4}$

13) $\frac{3}{4}$

14) $\frac{3}{12}$

15) $\frac{15}{20}$

16) $\frac{7}{25}$

17) $\frac{7}{50}$

18) $\frac{11}{50}$

19) $\frac{3}{20}$

20) $\frac{9}{20}$

Grade 8 Chapter 5 Lesson 8
Mental Math 15
Negative Fractions

1) $12(-\frac{5}{6})$

2) $-\frac{5}{9}\times 18$

3) $24(-\frac{3}{4})$

4) $-\frac{5}{8}\times 32$

5) $\frac{1}{2}\div(-\frac{1}{2})$

6) $\frac{1}{3}\div(-\frac{1}{6})$

7) $-\frac{1}{4}\div\frac{1}{12}$

8) $-2\div\frac{1}{2}$

9) $(-\frac{1}{2})^2$

10) $(-\frac{1}{5})^2$

11) $(-\frac{2}{5})^2$

12) $-(-\frac{3}{5})^2$

13) $-\frac{1}{4}-\frac{3}{4}$

14) $-\frac{1}{2}-\frac{3}{2}$

15) $-\frac{7}{8}+\frac{5}{8}$

16) $-\frac{7}{8}+\frac{3}{8}$

17) $-\frac{1}{2}-\frac{1}{4}$

18) $-\frac{3}{4}-\frac{1}{2}$

19) $-\frac{1}{2}+\frac{1}{4}$

20) $-\frac{3}{4}+\frac{1}{2}$

Grade 8 Chapter 5 Lesson 8
Mental Math 16
Negative Mixed Numbers

1) $-1-\frac{1}{2}$

2) $-\frac{1}{4}-2$

3) $-1+\frac{1}{2}$

4) $-2\frac{1}{2}+1$

5) $-2\frac{3}{8}-1\frac{1}{8}$

6) $-3\frac{3}{4}-2\frac{1}{4}$

7) $-4\frac{3}{4}+2\frac{1}{4}$

8) $-3\frac{5}{8}+2\frac{1}{8}$

9) $-3\frac{1}{2}-2\frac{1}{2}$

10) $-4\frac{1}{2}-2\frac{1}{2}$

11) $-2\frac{3}{4}+1\frac{1}{4}$

12) $-4\frac{5}{6}+1\frac{2}{6}$

13) $-2\frac{3}{8}+1\frac{1}{8}$

14) $-3\frac{3}{8}+2\frac{1}{8}$

15) $-4\frac{7}{8}+3\frac{3}{8}$

16) $-5\frac{5}{8}+4\frac{1}{8}$

17) $-2+1\frac{1}{2}$

18) $-2+1\frac{3}{4}$

19) $-3+1\frac{1}{4}$

20) $-3+1\frac{1}{5}$

Grade 8 Chapter 5 Lesson 9
Mental Math 17

Find Powers

1)	11^2	11)	21^2
2)	12^2	12)	22^2
3)	13^2	13)	23^2
4)	14^2	14)	24^2
5)	15^2	15)	25^2
6)	16^2	16)	26^2
7)	17^2	17)	27^2
8)	18^2	18)	28^2
9)	19^2	19)	29^2
10)	20^2	20)	30^2

Grade 8 Chapter 5 Lesson 9
Mental Math 18
Square Roots

1) $\sqrt{100}$
2) $\sqrt{121}$
3) $\sqrt{144}$
4) $\sqrt{169}$
5) $\sqrt{196}$
6) $\sqrt{225}$
7) $\sqrt{256}$
8) $\sqrt{289}$
9) $\sqrt{324}$
10) $\sqrt{361}$
11) $\sqrt{400}$
12) $\sqrt{441}$
13) $\sqrt{484}$
14) $\sqrt{529}$
15) $\sqrt{576}$
16) $\sqrt{625}$
17) $\sqrt{676}$
18) $\sqrt{729}$
19) $\sqrt{784}$
20) $\sqrt{841}$
21) $\sqrt{900}$
22) $\sqrt{961}$
23) $\sqrt{1600}$
24) $\sqrt{2500}$
25) $\sqrt{3600}$
26) $\sqrt{4900}$
27) $\sqrt{6400}$
28) $\sqrt{8100}$

Grade 8 Chapter 5 Lesson 9
Mental Math 19
Solve Equations

1) $x + 6 = 10$

2) $7 + x = 10$

3) $x - 9 = 5$

4) $x - 8 = 9$

5) $x - 7 = 8$

6) $12 - y = 3$

7) $13 - y = 8$

8) $14 - y = 5$

9) $15 - y = 7$

10) $16 - y = 9$

11) $8 \cdot x = 72$

12) $x \cdot 9 = 63$

13) $x \div 8 = 7$

14) $x \div 7 = 6$

15) $x \div 6 = 9$

16) $72 \div z = 9$

17) $72 \div z = 8$

18) $63 \div z = 9$

19) $56 \div z = 7$

20) $54 \div z = 9$

Grade 8 Chapter 5 Lesson 9
Mental Math 20
Proportions
Solve the proportions

$x \div 3 = 4 \div 2 \quad \frac{x}{3} = \frac{4}{2} \quad 2x = 12 \quad x = 6$

1) $\frac{1}{3} = \frac{x}{9}$

2) $\frac{4}{3} = \frac{20}{x}$

3) $\frac{9}{x} = \frac{3}{2}$

4) $\frac{3}{2} = \frac{15}{x}$

5) $\frac{3}{2} = \frac{x}{6}$

6) $\frac{12}{x} = \frac{3}{4}$

7) $\frac{x}{5} = \frac{18}{6}$

8) $\frac{x}{4} = \frac{6}{2}$

9) $\frac{9}{3} = \frac{6}{x}$

10) $\frac{44}{x} = \frac{8}{2}$

11) $y \div 6 = 4 \div 3$

12) $z \div 8 = 2 \div 4$

13) $y \div 3 = 6 \div 2$

14) $2 \div 8 = z \div 4$

15) $3 \div 6 = 2 \div y$

16) $8 \div 2 = z \div 4$

17) $y \div 3 = 12 \div 2$

18) $z \div 3 = 8 \div 4$

19) $y \div 4 = 4 \div 2$

20) $z \div 3 = 4 \div 2$

Grade 8 Chapter 5 Lesson 9

Mental Math 21
Law for Fractions

1) $\frac{1}{2}\times 6+\frac{1}{2}\times 4$

2) $\frac{4}{3}\times 9-\frac{1}{3}\times 9$

3) $\frac{1}{4}\times 4+\frac{3}{4}\times 4$

4) $\frac{6}{5}\times 15-\frac{1}{5}\times 15$

5) $\frac{1}{3}\times 6+\frac{2}{3}\times 6$

6) $\frac{3}{2}\times 16-\frac{1}{2}\times 16$

7) $\frac{1}{3}\times 12+\frac{2}{3}\times 12$

8) $\frac{5}{4}\times 8-\frac{1}{4}\times 8$

9) $\frac{1}{5}\times 10+\frac{4}{5}\times 10$

10) $\frac{7}{6}\times 12-\frac{1}{6}\times 12$

11) $\frac{1}{6}\times 12+\frac{5}{6}\times 12$

12) $\frac{5}{2}\times 8-\frac{1}{2}\times 8$

13) $\frac{3}{7}\times 14+\frac{4}{7}\times 14$

14) $\frac{7}{3}\times 6-\frac{1}{3}\times 6$

15) $\frac{3}{8}\times 16+\frac{5}{8}\times 16$

16) $\frac{5}{4}\times 4-\frac{1}{4}\times 4$

17) $\frac{4}{9}\times 18+\frac{5}{9}\times 18$

18) $\frac{7}{5}\times 10-\frac{2}{5}\times 10$

19) $\frac{3}{5}\times 15+\frac{2}{5}\times 15$

20) $\frac{7}{6}\times 12-\frac{1}{6}\times 12$

Grade 8 Chapter 5 Lesson 10
Reviews for Final Exam
Review 1 A

Date **Name**
Read directions carefully. **Answers must be in the indicated place.**
Show all steps for full credits. **Do only 1 operation per 1 step.**
Each problem is worth 5 points. **Solve 1 problem 5 minutes.**

Put - C before parentheses and change each sign inside parentheses

$-AC - AB = -C(A + B)$

$-7.5 \times 5 - 2.5 \times 5 = -5(7.5 + 2.5) = -5(10) = -50$

Simplify by laws for Negative Numbers

1) $-3.7 \times 4 - 1.3 \times 4$

2) $-4.9 \times 2 - 5.1 \times 2$

3) $-2.4 \times 3 - 1.6 \times 3$

4) $-9.6 \times 2 - 0.4 \times 2$

5) $-0.3 \times 6 - 0.7 \times 6$

6) $-0.7 \times 2 - 0.3 \times 2$

7) $-0.3 \times 8 - 0.3 \times 2$

8) $-3.6 \times 3 - 3.6 \times 7$

9) $-4.7 \times 9 - 5.3 \times 9$

10) $-2.4 \times 7 - 7.6 \times 7$

Grade 8 Chapter 5 Lesson 10
Review 1 B

Date
Name
Read directions carefully.
Answers must be in the indicated place.
Show all steps for full credits.
Do only 1 operation per 1 step.
Each problem is worth 5 points.
Solve 1 problem 5 minutes.

Put - C before parentheses and change each sign inside parentheses

$- AC + BC = - C (A - B)$

$- 12.5 (5) + 2.5(5) = - 5(12.5 - 2.5) = - 5(10) = - 50$

Simplify by Laws for Negative Numbers

1) $- 5.4 \times 5 + 3.4 \times 5$

2) $- 3.9 \times 6 + 1.9 \times 6$

3) $- 5.8 \times 3 + 1.8 \times 3$

4) $- 3.8 \times 4 + 1.8 \times 4$

5) $- 5.7 \times 8 + 4.7 \times 8$

6) $- 2.4 \times 6 + 2.4 \times 5$

7) $- 4.8 \times 3 + 4.8 \times 2$

8) $- 6.6 \times 8 + 1.6 \times 8$

9) $- 6.4 \times 3 + 1.4 \times 3$

10) $- 6.2 \times 4 + 5.2 \times 4$

Grade 8 Chapter 5 Lesson 10
Review 1 C

Date	**Name**
Read directions carefully.	**Answers must be in the indicated place.**
Show all steps for full credits.	**Do only 1 operation per 1 step.**
Each problem is worth 5 points.	**Solve 1 problem 5 minutes.**

To solve problems work by order of operations or change division for multiplication then put C before parentheses.

AC + BC = C (A + B) **AC – B C = C (A - B)**

a) $2\frac{5}{8} \div \frac{1}{8} + 1\frac{3}{8} \div \frac{1}{8} = 2\frac{5}{8} \times 8 + 1\frac{3}{8} \times 8 = 8(2\frac{5}{8} + 1\frac{3}{8}) = 8(4) = 32$

b) $2\frac{5}{8} \div \frac{1}{8} - 1\frac{3}{8} \div \frac{1}{8} = 2\frac{5}{8} \times 8 - 1\frac{3}{8} \times 8 = 8(2\frac{5}{8} - 1\frac{3}{8}) = 8(1\frac{2}{8}) = 8(\frac{10}{8}) = 10$

Simplify by Laws for Fractions

1) $1\frac{3}{4} \div \frac{1}{4} + 1\frac{1}{4} \div \frac{1}{4}$

2) $4\frac{3}{5} \div \frac{1}{5} + 1\frac{2}{5} \div \frac{1}{5}$

3) $2\frac{1}{2} \div \frac{1}{2} + 1\frac{1}{2} \div \frac{1}{2}$

4) $1\frac{5}{6} \div \frac{1}{6} + 1\frac{1}{6} \div \frac{1}{6}$

5) $1\frac{2}{3} \div \frac{1}{3} + 1\frac{1}{3} \div \frac{1}{3}$

6) $3\frac{1}{2} \div \frac{1}{2} - 1\frac{1}{2} \div \frac{1}{2}$

7) $1\frac{3}{4} \div \frac{1}{4} - 1\frac{1}{4} \div \frac{1}{4}$

8) $3\frac{2}{3} \div \frac{1}{3} - 1\frac{2}{3} \div \frac{1}{3}$

9) $4\frac{3}{5} \div \frac{1}{5} - 1\frac{3}{5} \div \frac{1}{5}$

10) $4\frac{5}{6} \div \frac{1}{6} - 2\frac{1}{6} \div \frac{1}{6}$

Grade 8 Chapter 5 Lesson 10
Review 1 D

Date **Name**

Read directions carefully. **Answers must be in the indicated place.**

Show all steps for full credits. **Do only 1 operation per 1 step.**

Each problem is worth 5 points. **Solve 1 problem 5 minutes.**

To solve problems work by order of operations or change division for multiplication. Put - C before parentheses and change each sign inside parentheses. $-AC - BC = -C(A + B)$ $-AC + BC = -C(A - B)$

a) $-2\frac{5}{8} \div \frac{1}{8} - 1\frac{3}{8} \div \frac{1}{8} = -2\frac{5}{8} \times 8 - 1\frac{3}{8} \times 8 = -8(2\frac{5}{8} + 1\frac{3}{8}) = -8(4) = -32$

b) $-2\frac{5}{8} \div \frac{1}{8} + 1\frac{3}{8} \div \frac{1}{8} = -2\frac{5}{8} \times 8 + 1\frac{3}{8} \times 8 = -8(2\frac{5}{8} - 1\frac{3}{8}) = -8(1\frac{2}{8}) = -8(\frac{10}{8}) = -10$

Simplify by Laws for Fractions

1) $-1\frac{3}{4} \div \frac{1}{4} - 1\frac{1}{4} \div \frac{1}{4}$

2) $-4\frac{3}{5} \div \frac{1}{5} - 1\frac{2}{5} \div \frac{1}{5}$

3) $-2\frac{1}{2} \div \frac{1}{2} - 1\frac{1}{2} \div \frac{1}{2}$

4) $-1\frac{5}{6} \div \frac{1}{6} - 1\frac{1}{6} \div \frac{1}{6}$

5) $-1\frac{2}{3} \div \frac{1}{3} - 1\frac{1}{3} \div \frac{1}{3}$

6) $-3\frac{1}{2} \div \frac{1}{2} + 1\frac{1}{2} \div \frac{1}{2}$

7) $-1\frac{3}{4} \div \frac{1}{4} + 1\frac{1}{4} \div \frac{1}{4}$

8) $-3\frac{2}{3} \div \frac{1}{3} + 1\frac{2}{3} \div \frac{1}{3}$

9) $-4\frac{3}{5} \div \frac{1}{5} + 1\frac{3}{5} \div \frac{1}{5}$

10) $-4\frac{5}{6} \div \frac{1}{6} + 2\frac{1}{6} \div \frac{1}{6}$

Grade 8 Chapter 5 Lesson 11
Laws for operations

Date | **Name**

Read directions carefully. | **Answers must be in the indicated place.**

Show all steps for full credits. | **Do only 1 operation per 1 step.**

Each problem is worth 5 points. | **Solve 1 problem 5 minutes.**

Distributive laws of multiplication

(A + B) C = AC + BC | **(A - B) C = AC – B C**

Example a) $-\frac{5}{6}\times 3 = -\frac{5}{6}\times\frac{3}{1} = -\frac{5}{2} = -2\frac{1}{2} = -2.5$

Example b) $\frac{5}{3}\times(-6) = -6\times\frac{5}{3} = -6\div 3\times 5 = -2\times 5 = -10$

Example c) $\left(-1\frac{1}{3}+\frac{5}{6}\right)\times 3 = -\frac{4}{3}\times\frac{3}{1}+\frac{5}{6}\times\frac{3}{1} = -4+\frac{5}{2} = -4+2.5 = -1.5$

Worksheet 7 A
Evaluate by using laws

1) $9(-1+\frac{4}{9})$

2) $(-\frac{2}{3}+\frac{3}{5})\times 15$

3) $9(-3+\frac{5}{9})$

4) $(-1-\frac{5}{6})\times 6$

5) $7(-2-\frac{3}{7})$

6) $(-1+\frac{1}{6})\times 6$

7) $6(-1\frac{1}{2}+\frac{2}{3})$

8) $(-\frac{5}{7}-\frac{4}{7})\times 7$

9) $7(-\frac{5}{7}+\frac{4}{7})$

10) $(-3-\frac{2}{3})\times 3$

11) $3(-2-\frac{1}{6})$

12) $(-2\frac{1}{4}+\frac{5}{8})\times 4$

13) $3(-1\frac{5}{6}+1\frac{1}{3})$

14) $(-3\frac{1}{2}+\frac{1}{4})\times 4$

15) $3(-1\frac{2}{3}+\frac{5}{6})$

16) $(-1\frac{2}{7}+\frac{1}{3})\times 21$

17) $6(-2\frac{1}{2}-1\frac{1}{3})$

18) $(-4-\frac{5}{8})\times 2$

19) $3(-1\frac{1}{3}+\frac{5}{12})$

20) $(-1\frac{1}{4}+\frac{5}{8})\times 4$

Grade 8 Chapter 5 Lesson 11

Solve by finding common factor

Put C before parentheses

- AC - BC = - C (A + B)   - AC + B C = - C (A - B)

Example $-1\frac{1}{4}\times\frac{3}{4}-\frac{3}{4}\times2\frac{3}{4}=-\frac{3}{4}(1\frac{1}{4}+2\frac{3}{4})=-\frac{3}{4}\times\frac{4}{1}=-3$

Worksheet 7 B

Evaluate by using Common Factor

1) $-8\times1\frac{3}{4}+8\times\frac{5}{8}$

2) $-3\frac{2}{9}\times\frac{3}{5}-\frac{3}{5}\times6\frac{7}{9}$

3) $-7\times1\frac{1}{7}+7\times\frac{5}{14}$

4) $-4\times2\frac{1}{4}+4\times1\frac{7}{8}$

5) $-2\frac{6}{7}\times\frac{7}{10}-2\frac{1}{7}\times\frac{7}{10}$

6) $-5\frac{5}{6}\times\frac{3}{4}+1\frac{5}{6}\times\frac{3}{4}$

7) $-4\frac{4}{5}\times\frac{5}{8}-\frac{5}{8}\times3\frac{1}{5}$

8) $-3\frac{2}{3}\times1\frac{1}{2}+1\frac{2}{3}\times1\frac{1}{2}$

9) $-\frac{5}{6}\times5\frac{1}{5}+\frac{5}{6}\times2\frac{1}{5}$

10) $-3\frac{1}{3}\times1\frac{1}{5}-1\frac{1}{5}\times1\frac{2}{3}$

11) $-1\frac{3}{5}\times2\frac{1}{2}-1\frac{3}{5}\times7\frac{1}{2}$

12) $-4\frac{2}{3}\times\frac{3}{4}+3\frac{1}{3}\times\frac{3}{4}$

13) $-4\frac{3}{5}\times\frac{5}{6}+\frac{5}{6}\times1\frac{3}{5}$

14) $-4\frac{1}{2}\times\frac{1}{6}-\frac{1}{6}\times1\frac{1}{2}$

15) $-3\frac{3}{4}\times1\frac{1}{6}+2\frac{1}{4}\times1\frac{1}{6}$

16) $-\frac{1}{3}\times15+\frac{1}{5}\times15$

17) $-1\frac{3}{4}\times8+1\frac{1}{2}\times8$

18) $-1\frac{1}{2}\times6+1\frac{1}{6}\times6$

19) $-2\frac{1}{2}\times6+1\frac{1}{2}\times6$

20) $-1\frac{3}{4}\times12+1\frac{1}{4}\times12$

21) $(-1\frac{3}{5}-1\frac{2}{5})\div\frac{1}{5}$

22) $(-1\frac{3}{5}+1\frac{2}{5})\div\frac{2}{5}$

23) $(-4\frac{3}{5}-1\frac{2}{5})\div\frac{1}{5}$

24) $(-3\frac{2}{3}+1\frac{2}{3})\div\frac{2}{3}$

25) $(-1\frac{3}{4}+1\frac{1}{2})\div\frac{1}{8}$

26) $(-2\frac{1}{2}-1\frac{1}{2})\div\frac{1}{6}$

27) $(-1\frac{4}{5}-1\frac{1}{5})\div\frac{1}{5}$

28) $(-4\frac{3}{5}+1\frac{3}{5})\div\frac{1}{5}$

29) $(-1\frac{1}{3}-1\frac{2}{3})\div\frac{1}{3}$

30) $(-4\frac{1}{5}+2\frac{1}{5})\div\frac{1}{5}$

Grade 8 Chapter 5 Lesson 12
Proportions

Date | **Name**

Read directions carefully. **Answers must be in the indicated place.**
Show all steps for full credits. **Do only 1 operation per 1 step.**
Each problem is worth 5 points. **Solve 1 problem 5 minutes.**

Solve the proportions

$$x \div 3 = 4 \div 2 \qquad \frac{x}{3} = \frac{4}{2} \qquad 2x = 12 \qquad x = 6$$

Worksheet 8 A
Solve the proportions

1) $\frac{24}{72} = \frac{x}{9}$

2) $\frac{60}{45} = \frac{24}{x}$

3) $\frac{42}{x} = \frac{51}{34}$

4) $\frac{18}{12} = \frac{75}{x}$

5) $\frac{69}{46} = \frac{x}{54}$

6) $\frac{4.2}{x} = \frac{5.7}{3.8}$

7) $\frac{x}{5.1} = \frac{2.7}{0.9}$

8) $\frac{x}{4.7} = \frac{27.3}{9.1}$

9) $\frac{7.8}{2.6} = \frac{4.5}{x}$

10) $\frac{4.4}{x} = \frac{8.8}{2.2}$

11) $\frac{y}{5\frac{2}{3}} = \frac{1\frac{7}{9}}{1\frac{7}{27}}$

12) $\frac{z}{\frac{4}{5}} = \frac{3\frac{1}{8}}{1\frac{1}{4}}$

13) $y \div \frac{3}{5} = 3\frac{1}{8} \div 3\frac{3}{4}$

14) $1\frac{7}{8} \div 8\frac{1}{3} = z \div \frac{4}{9}$

15) $3\frac{3}{4} \div 1\frac{1}{8} = 2\frac{2}{3} \div y$

16) $\frac{3}{8} \div \frac{3}{11} = z \div \frac{4}{11}$

17) $y \div \frac{1}{3} = \frac{3}{4} \div \frac{1}{2}$

18) $z \div \frac{3}{14} = 3\frac{1}{9} \div \frac{4}{9}$

19) $y \div 4\frac{1}{6} = \frac{4}{7} \div \frac{20}{21}$

20) $z \div 3\frac{1}{5} = 4\frac{1}{2} \div 2\frac{1}{4}$

Grade 8 Chapter 5 Lesson 12
Proportions

Date | **Name**

Read directions carefully. **Answers must be in the indicated place.**
Show all steps for full credits. **Do only 1 operation per 1 step.**
Each problem is worth 5 points. **Solve 1 problem 5 minutes.**

Example: Solve the proportions

$3\frac{1}{3} \div 2\frac{1}{2} = 2\frac{2}{3} \div x$ $\quad \frac{3\frac{1}{3}}{2\frac{1}{2}} = \frac{2\frac{2}{3}}{x}$ $\quad 3\frac{1}{3}x = 2\frac{1}{2} \cdot 2\frac{2}{3}$ $\quad x = 2\frac{1}{2} \cdot 2\frac{2}{3} \div 3\frac{1}{3}$

$x = \frac{5}{2} \cdot \frac{8}{3} \div \frac{10}{3}$ $\quad x = \frac{5}{2} \cdot \frac{8}{3} \cdot \frac{3}{10}$ $\quad x = \frac{1 \cdot 4}{1 \cdot 2}$ $\quad x = 2$

Worksheet 8 B
Solve the proportions

21) $\frac{3}{14} \div 3\frac{1}{7} = y \div 3\frac{2}{3}$

22) $1\frac{1}{2} \div y = 4\frac{3}{4} \div 2\frac{3}{8}$

23) $y \div \frac{4}{25} = \frac{3}{4} \div \frac{4}{5}$

24) $y \div 2\frac{1}{3} = \frac{4}{7} \div 4\frac{1}{6}$

25) $7\frac{1}{3} \div 2\frac{1}{2} = 3\frac{2}{3} \div y$

26) $6\frac{1}{2} \div 3\frac{1}{4} = 3\frac{1}{2} \div k$

27) $m \div 2\frac{1}{2} = 3\frac{1}{4} \div 6\frac{1}{2}$

28) $n \div 2.7 = 8\frac{1}{3} \div 0.5$

29) $2.8 \div p = 1\frac{2}{3} \div 2\frac{6}{7}$

30) $d \div 6.4 = 1\frac{7}{8} \div 7\frac{1}{2}$

31) $c \div 3\frac{1}{3} = 0.9 \div 0.4$

32) $3\frac{1}{3} \div d = 3\frac{3}{5} \div 2.7$

33) $1\frac{7}{9} \div g = 9\frac{1}{3} \div 2\frac{5}{8}$

34) $h \div \frac{7}{12} = \frac{2}{3} \div \frac{7}{9}$

35) $3\frac{3}{5} \div k = 1\frac{4}{5} \div 1\frac{1}{4}$

36) $2\frac{1}{2} \div 1\frac{1}{2} = 6\frac{2}{3} \div m$

37) $3\frac{3}{5} \div \frac{4}{15} = n \div \frac{5}{9}$

38) $2 \div p = 1\frac{4}{7} \div 3\frac{1}{7}$

39) $5\frac{1}{4} \div 3\frac{1}{2} = 4\frac{1}{2} \div u$

40) $3\frac{1}{3} \div 2\frac{1}{2} = 2\frac{2}{3} \div w$

Grade 8 Chapter 5 Lesson 13
College Placement Test 8 A (50 minutes)
(Basic Problems)

Date **Name**

Read directions carefully. **Answers must be in the indicated place.**

Show all steps for full credits. **Do only 1 operation per 1 step.**

Each problem is worth 5 points. **Solve 1 problem 5 minutes.**

1) $39600 \div (85^2 - 25^2)$

2) $4.9 \times 7 - 1.8 \times 7 + 6.9 \times 7$

3) $2\frac{1}{2} \times 16 - 1\frac{1}{4} \times 16$

4) $-2\frac{6}{7} \div 1\frac{3}{7} + 2\frac{1}{7} \div 1\frac{3}{7}$

5) Simplify $4(\frac{1}{2}c - \frac{1}{4}d) - \frac{2}{3}(6c + 3d)$

6) Solve the equation $0.3y + 0.2(2y-3) = 0.8 - 0.7(y-2)$

Solve the word problems

7) Find the perimeter of right triangle if hypotenuse is $5\frac{1}{10}$ ft and leg is $2\frac{2}{5}$ ft.

8) Find the area of trapezoid, if height is $4\frac{2}{3}$ m, bases are $8\frac{1}{2}$ m and $6\frac{1}{2}$ m.

9) A bus drove 75 mi at a speed of 30 mi/h.
How much time it spent?

10) Of 32 children, 62.5% were boys. How many were girls?

Grade 8 Chapter 5 Lesson 13
College Placement Test 8 B (50 minutes)
(Basic Problems)

Date **Name**

Read directions carefully. **Answers must be in the indicated place.**

Show all steps for full credits. **Do only 1 operation per 1 step.**

Each problem is worth 5 points. **Solve 1 problem 5 minutes.**

Perform the indicated operations.

1) $70200 \div (95^2 - 35^2)$

2) $3.4 \times 5 - 1.7 \times 5 + 8.3 \times 5$

3) $1\frac{3}{4} \times 12 - 1\frac{1}{4} \times 12$

4) $-4\frac{4}{5} \div 1\frac{3}{5} + 3\frac{1}{5} \div 1\frac{3}{5}$

5) Simplify $16(\frac{3}{4}x + \frac{1}{4}y) - 15(\frac{2}{3}x + \frac{1}{3}y)$

6) Solve the equation $-5(4.2y + 1) + 4(1.4y - 2) = -20.7$

7) Find the perimeter of right triangle if hypotenuse is $7\frac{4}{5}$ ft and leg is 3 ft.

8) Find the area of trapezoid, if it height is $2\frac{2}{3}$ m, bases are $7\frac{1}{2}$ m and $4\frac{1}{2}$ m.

9) A bus drove 36 mi at a speed of 15 mi/h.
How much time it spent?

10) Of 24 children, 37.5% were boys. How many were girls?

Grade 8 Chapter 5 Lesson 14
Final Exam 8 A
(Basic Problems)

Date | **Name**

Read directions carefully. | **Answers must be in the indicated place.**

Show all steps for full credits. | **Do only 1 operation per 1 step.**

Each problem is worth 5 points. | **Solve 1 problem 5 minutes.**

Always try to find the easiest way to calculate, that way, fewer mistakes will made.

Perform the indicated operations

1) $3^2 - 0.9^2 \div 0.3^2$

1) ______

2) $\left(\frac{1}{4} - \frac{1}{8}\right)^2 \div \frac{1}{32}$

2) ______

3) $2\frac{3}{4} \div \frac{1}{16} - 1\frac{1}{4} \div \frac{1}{16}$

3) ______

4) $(-\frac{2}{5})^2 - (-\frac{3}{5})^2$

4) ______

5) $-\frac{5}{6} \times 5\frac{1}{5} + \frac{5}{6} \times 2\frac{1}{5}$

5) ---

Grade 8 Chapter 5 Lesson 14
Final Exam 8 A
(Basic Problems)

Date	**Name**
Read directions carefully.	**Answers must be in the indicated place.**
Show all steps for full credits.	**Do only 1 operation per 1 step.**
Each problem is worth 5 points.	**Solve 1 problem 5 minutes.**

Simplify expression

6) $-6(\frac{1}{6}-\frac{1}{2}x)-2(1-\frac{1}{2}x)$

6)_____

Solve the equation

7) $0.3x + 0.2\ (2x - 3) = 0.8 - 0.7\ (x - 2)$

7) _____

Solve the word problems

8) Find area of trapezoid, if height is $5\frac{1}{3}$ ft, bases are $6\frac{1}{2}$ ft and $5\frac{1}{2}$ ft

8) _____

9) What speed does a boy have, if he drove 57 mi in 4 hours?

9) ______

10) A factory has 5 new machines and 80% were old machine.
How many machines are in the factory?

10)_____

Grade 8 Chapter 5 Lesson 14
Final Exam 8 B (Basic Problems)

Date **Name**

Read directions carefully. **Answers must be in the indicated place.**

Show all steps for full credits. **Do only 1 operation per 1 step.**

Each problem is worth 5 points. **Solve 1 problem 5 minutes.**

Always try to find the easiest way to calculate, that way, fewer mistakes will made.

Perform the indicated operations

1) $2^2 - 0.4^2 \div 0.2^2$ 1) _____

2) $\left(\frac{1}{5} - \frac{1}{10}\right)^2 \div \frac{1}{10}$ 2) _____

3) $1\frac{3}{4} \div \frac{1}{12} - 1\frac{1}{4} \div \frac{1}{12}$ 3) _____

4) $-(-\frac{4}{5})^2 + (-\frac{1}{5})^2$ 4) _____

5) $-3\frac{1}{3} \times 1\frac{1}{5} - 1\frac{1}{5} \times 1\frac{2}{3}$ 5) _____

Grade 8 Chapter 5 Lesson 14
Final Exam 8 B (Basic Problems)

Date | **Name**

Read directions carefully. | **Answers must be in the indicated place.**

Show all steps for full credits. | **Do only 1 operation per 1 step.**

Each problem is worth 5 points. | **Solve 1 problem 5 minutes.**

Always try to find the easiest way to calculate, that way, fewer mistakes will made.

Simplify expression

6) $-8(\frac{1}{2}-\frac{1}{4}y)-3(\frac{2}{3}-\frac{1}{3}y)$ 6)_____

Solve the equation

7) $-2(3.1x-1)+3(1.2x+1)=-14.5$ 7) _____

Solve the word problems

8) Find area of trapezoid, if height is $2\frac{2}{3}$ ft, bases are $7\frac{1}{2}$ ft and $4\frac{1}{2}$ ft 8) _____

9) What speed does a boy have, if he drove 73 mi in 4 hours? 9) ______

10) A factory has 10 new machines and 60% were old machine. How many machines are in the factory? 10)_______

Grading Scale: A =10 points, B=9points, C=8 points, D=7 points, F=0-6 points

Lesson 14 Final Exam 8 B (Basic Problems)

1) 0	2) 0.1	3) 6	4) -0.6	5) -6
6) 2	7) 7.5	8) 16 cm^2	9) 18.25 mi/hr.	10) 25 machines

Grade 8 Answers Chapter 5

Perfect Square Numbers

Take number 20 from the left column add number 5 from the first row to get number 25. Answer is in intersection row of 20 and column of 5 $25^2 = 25 \cdot 25 = 625$

Memorize Table of Perfect Square for Indicated Numbers

a^2	0	1	2	3	4	5	6	7	8	9
	0	1	4	9	16	25	36	49	64	81
10	100	121	144	169	196	225	256	289	324	361
20	400	441	484	529	576	625	676	729	784	841
30	900	961	1024	1089	1156	1225	1296	1369	1444	1521
40	1600	1681	1764	1849	1936	2025	2116	2209	2304	2401
50	2500	2601	2704	2809	2916	3025	3136	3249	3364	3481
60	3600	3721	3844	3969	4096	4225	4356	4489	4624	4761
70	4900	5041	5184	5329	5476	5625	5776	5929	6084	6241
80	6400	6561	6724	6889	7056	7225	7396	7569	7744	7921
90	8100	8281	8464	8649	8836	9025	9216	9409	9604	9801

Square Roots from Perfect Square Numbers

The square root of three and four-digit perfect square numbers is always a two-digit number. Divide the number under the root by groups of two digits, from right to left. The last group can be one or two digits. To find the first digit, find the number whose square is smaller than or equal to the last group. To find the second digit, use the rule:

Last digit of given number	1	4	5	6	9
Last digit of answer	1 or 9	2 or 8	5	4 or 6	3 or 7

$31^2 = 31 \times 31 = 961$ $\quad$ $32^2 = 32 \times 32 = 1024$

$\sqrt{9'61} = 31$ $\quad$ $\sqrt{10'24} = 32$

Grade 8 Answers Chapter 5

Lesson 7 Mental Math 6

1)30 2) 25 3) 40 4) 4 5) 50 6) 4 7) 50 8) 3 9) 30 10) 5
11) 800 12) 60 13) 60 14) 160 15) 150 16) 80 17) 120 18) 60 19) 150 20) 40

Lesson 7 Mental Math 7

1) 1 2) 8 3) 27 4) 64 5) 125 6) 216 7) 343 8) 512 9) 729 10) 1,000
11) 2 12) 4 13) 8 14) 16 15) 32 16) 64 17) 128 18) 256 19) 512 20) 1024

Lesson 7 Mental Math 8

1) 2^2 2) 3^2 3) 2^4 4) 5^2 5) $2^2 \cdot 3^2$ 6) 7^2 7) 2^6 8) 3^4 9) $2 \cdot 3$ 10) 2^3 11) $2 \cdot 5$
12) $2^2 \cdot 3$ 13) $2 \cdot 7$ 14) $3 \cdot 5$ 15) $2 \cdot 3^2$ 16) $2^2 \cdot 5$ 17) $2^3 \cdot 3$ 18) 3^3 19) $2 \cdot 3 \cdot 5$ 20) 2^5

Lesson 7 Mental Math 9

1) 0.2 2) 1.5 3) 0.7 4) 0.7 5) 2 6) 0.6 7) 0 8) 0.3
9) 0 10) 0 11) 1.9 12) 0.8 13) 1 14) 2 15) 3 16) 4
17) 6 18) 7 19) 0 20) $\emptyset$ 21) 0.8 22) 1 23) 0.04 24) 0.09
25) 0.16 26) 0.25 27) 0.36 28) 0.49 29) 0.64 30) 0.81

Lesson 7 Mental Math 10

1) 80 2) 20 3) 4 4) 10 5) 70 6) 20 7) 24 8) 16 9) 15 10) 12
11) 8 12) 30 13) 24 14) 14 15) 5 16) 8 17) 18 18) 6 19) 8 20) 8

Lesson 7 Mental Math 11

1) $\frac{1}{5}$ 2) $\frac{1}{4}$ 3) 5 4) 1 5) 1 6) 2 7) 3 8) 4 9) $\frac{1}{4}$ 10) $\frac{1}{25}$
11) $\frac{4}{25}$ 12) $\frac{9}{25}$ 13) 1 14) 2 15) $\frac{1}{4}$ 16) $\frac{1}{2}$ 17) $\frac{3}{4}$ 18) $1\frac{1}{4}$ 19) $\frac{1}{4}$ 20) $\frac{1}{4}$

Lesson 7 Mental Math 12

1) $1\frac{1}{2}$ 2) $2\frac{1}{4}$ 3) $\frac{1}{2}$ 4) $1\frac{1}{2}$ 5) $3\frac{1}{2}$ 6) 6 7) $2\frac{1}{2}$ 8) $1\frac{1}{2}$ 9) 6 10) 7
11) $1\frac{1}{2}$ 12) $3\frac{1}{2}$ 13) $1\frac{1}{4}$ 14) $1\frac{1}{4}$ 15) $1\frac{1}{2}$ 16) $1\frac{1}{2}$ 17) $\frac{1}{2}$ 18) $\frac{1}{4}$ 19) $1\frac{3}{4}$ 20) $1\frac{4}{5}$

Lesson 8 Mental Math 13

1) 5 2) 6 3) 8 4) 5 5) 9 6) 8 7) 6 8) 4 9) 5 10) 6
11) 6 12) 8 13) 7 14) 6 15) 8 16) 8 17) 9 18) 5 19) 5 20) 6

Lesson 8 Mental Meth 14

1) 0.3 2) 0.09 3) 0.007 4) 0.5 5) 0.5 6) 0.2 7) 0.4 8) 0.6 9) 0.8 10) 0.8
11)0.25 12)0.5 13)0.75 14) 0.25 15) 0.75 16) 0.28 17) 0.14 18) 0.22 19) 0.15 20) 0.45

Lesson 8 Mental Math 15

1) - 10 2) -10 3) – 18 4) – 20 5) – 1 6) -2 7) – 3 8) - 4 9) $\frac{1}{4}$ 10) $\frac{1}{25}$ 11) $\frac{4}{25}$
12) $-\frac{9}{25}$ 13) – 1 14) -2 15) $-\frac{1}{4}$ 16) $-\frac{1}{2}$ 17) $-\frac{3}{4}$ 18) $-1\frac{1}{4}$ 19) $-\frac{1}{4}$ 20) $-\frac{1}{4}$

Grade 8 Answers Chapter 5

Lesson 8 Mental Math 16

1) $-1\frac{1}{2}$ 2) $-2\frac{1}{4}$ 3) $-\frac{1}{2}$ 4) $-1\frac{1}{2}$ 5) $-3\frac{1}{2}$ 6) – 6 7) $-2\frac{1}{2}$ 8) $-1\frac{1}{2}$ 9) – 6 10) – 7

11) $-1\frac{1}{2}$ 12) $-3\frac{1}{2}$ 13) $-1\frac{1}{4}$ 14) $-1\frac{1}{4}$ 15) $-1\frac{1}{2}$ 16) $-1\frac{1}{2}$ 17) $-\frac{1}{2}$ 18) $-\frac{1}{4}$ 19) $-1\frac{3}{4}$ 20) $-1\frac{4}{5}$

Lesson 9 Mental Math 17

1) 121 2) 144 3) 169 4) 196 5) 225 6) 256 7) 289 8) 324 9) 361 10) 400
11) 441 12) 484 13) 529 14) 576 15) 625 16) 676 17) 729 18) 784 19) 841 20) 900

Lesson 9 Mental Math 18

1) 10 2) 11 3) 12 4) 13 5) 14 6) 15 7) 16 8) 17 9) 18 10) 31
11) 20 12) 21 13) 22 14) 23 15) 24 16) 25 17) 26 18) 27 19) 28 20) 29
21) 30 22) 31 23) 40 24) 50 25) 60 26) 70 27) 80 28) 90

Lesson 9 Mental Math 19

1) 4 2) 3 3) 14 4) 17 5) 15 6) 9 7) 5 8) 9 9) 8 10) 7
11) 9 12) 7 13) 56 14) 42 15) 54 16) 8 17) 9 18) 7 19) 8 20) 6

Lesson 9 Mental Math 20

1) 3 2) 15 3) 6 4) 10 5) 9 6) 16 7) 15 8) 12 9) 2 10) 11
11) 8 12) 4 13) 9 14) 1 15) 4 16) 16 17) 18 18) 6 19) 8 20) 6

Lesson 9 Mental Math 21

1)5 2) 9 3) 4 4) 15 5) 6 6) 16 7) 12 8) 8 9) 10 10) 12
11) 12 12) 16 13) 14 14) 12 15) 16 16) 4 17) 18 18) 10 19) 15 20) 12

Lesson 10 Review for Final Exam 1 Review 1 A

1) - 20 2) - 20 3) - 12 4) - 20 5) - 6
6) - 2 7) - 3 8) - 36 9) - 90 10) - 70

Review 1 B 1) - 10 2) - 12 3) - 12 4) - 8 5) - 8
6) - 2.4 7) - 4.8 8) - 40 9) - 15 10) - 4

Review 1 C

1) 12 2) 30 3) 8 4) 18 5) 9
6) 4 7) 2 8) 6 9) 15 10) 16

Review 1 D

1) - 12 2) - 30 3) - 8 4) - 18 5) - 9
6) - 4 7) - 2 8) - 6 9) - 15 10) - 16

Grade 8 Answers Chapter 5

Lesson 11 Worksheet 7 A

1) -5	2) -1	3) -22	4) -11	5) -17
6) -5	7) -5	8) -9	9) -1	10) -11
11) -6.5	12) -6.5	13) -1.5	14) - 13	15) -2.5
16) -20	17) -23	18) -9.25	19) -2.75	20) -2.5

Lesson 11 Worksheet 7 B

1) -9	2) -6	3) -5.5	4) -1.5	5) -3.5
6) - 3	7) -5	8) -3	9) -2.5	10) -6
11) - 16	12) - 1	13) - 2.5	14) - 1	15) -1.75
16) - 2	17) - 2	18) - 2	19) - 6	20) - 6
21) - 15	22) - 0.5	23) - 30	24) - 3	25) - 2
26) - 24	27) - 15	28) - 15	29) - 9	30) - 10

Lesson 12 Worksheet 8 A

1) 3	2) 18	3) 28	4) 50	5) 81
6) 2.8	7) 15.3	8) 14.1	9) 1.5	10) 1.1
11) 8	12) 2	13) 0.5	14) 0.1	15) 0.8
16) 0.5	17) 0.5	18) 1.5	19) 2.5	20) 6.4

Worksheet 8 B

21) 0.25	22) 0.75	23) 0.15	24) 0.32	25) 1.25
26) 1.75	27) 1.25	28) 45	29) 4.8	30) 1.6
31) 7.5	32) 2.5	33) 0.5	34) 0.5	35) 2.5
36) 4	37) 7.5	38) 4	39) 3	40) 2

Lesson 13 College Placement Test 8 A (Basic Problems)

1) 6 2) 70 3) 20 4) – 0.5 5) -2c – 3d 6) 2 7) 12 ft 8) 35 in^2 9) 2.5 hr. 10) 12 girls

Lesson 13 College Placement Test 8 B (Basic Problems)

1) 9 2) 50 3) 6 4) – 1 5) 2x – y 6) 0.5 7) 18 ft 8) 16m^2 9) 2.4 hr. 10) 15 girls

Lesson 14 Final Exam 8 A (Basic Problems)

1) 0	2) 0.5	3) 24	4) -0.2	5) -2.5
6) 0.75	7) 2	8) 32 ft^2	9) 14.25 mi/hr.	10) 25 machines

Lesson 14 Final Exam 8 B (Basic Problems)

1) 0	2) 0.1	3) 6	4) -0.6	5) -6
6) 2	7) 7. 5	8) 16 cm^2	9) 18.25 mi/hr.	10) 25 machines

Grade 8

Chapter 6

Extra

Reviews

Grade 8 Chapter 6 Contents
Extra Reviews
Lesson 1 – 10
Basic Problems

Memorize Exponents
(for good memory)

1)	$1^2=$	11)	$1^3=$	21)	$2^1=$
2)	$2^2=$	12)	$2^3=$	22)	$2^2=$
3)	$3^2=$	13)	$3^3=$	23)	$2^3=$
4)	$4^2=$	14)	$4^3=$	24)	$2^4=$
5)	$5^2=$	15)	$5^3=$	25)	$2^5=$
6)	$6^2=$	16)	$6^3=$	26)	$2^6=$
7)	$7^2=$	17)	$7^3=$	27)	$2^7=$
8)	$8^2=$	18)	$8^3=$	28)	$2^8=$
9)	$9^2=$	19)	$9^3=$	29)	$2^9=$
10)	$10^2=$	20)	$10^3=$	30)	$2^{10}=$

Grade 8 Chapter 6 Lesson 1
College Placement Test (50 minutes)
Evaluate

1) $18000 \div (75^2 - 45^2)$

2) $5.6 \times 4 + 2.7 \times 4 - 8.3 \times 4$

3) $3\frac{3}{5} \times 15 - 2\frac{1}{5} \times 15$

4) $-3\frac{2}{3} \div \frac{5}{6} + 1\frac{1}{3} \div \frac{5}{6}$

5) Simplify $\frac{3}{7}(14x - 21y) - 15(\frac{2}{3}x - \frac{1}{3}y)$

6) Solve the equation $-2(3.1x - 1) + 3(1.2x + 1) = -14.5$

7) Find the perimeter of right triangle if it hypotenuse is $8\frac{1}{2}$ ft and another leg is 1.3 ft.

8) Find the area of trapezoid, if it height is $6\frac{2}{3}$ in, bases are $8\frac{1}{2}$ in and $9\frac{1}{2}$ in.

9) A bus drove 48 mi at a speed of 20 mi/h.
How much time bus spend?

10) Of 16 children, 12.5% were boys. How many were girls?

How you feel about Math:

a) Love b) Like c) Nothing d) Do not like e) Hate

Grade 8 Chapter 6 Lesson 1
Whole Numbers
Review 1 A

Date | **Name**

Read directions carefully.
Show all steps for full credits.
Each problem is worth 5 points.
Answers must be in the indicated place.
Do only 1 operation per 1 step.
Solve 1 problem 5 minutes.

Evaluate

1) $450607 + 506078$

2) $900000 - 887766$

3) 908×605

4) 4506×809

5) $686588 \div 98$

6) $641048 \div 908$

7) $363600 \div 80 \times 9$

8) $8100 - 7263 \div 9$

9) $500 - 320 \div 8 \times 5$

10) $150 \div 5 \times 3 - 120 \div 4 \times 3$

Grade 8 Chapter 6 Lesson 1
Decimals
Review 2 A

Date | **Name**

Read directions carefully.
Show all steps for full credits.
Each problem is worth 5 points.

Answers must be in the indicated place.
Do only 1 operation per 1 step.
Solve 1 problem 5 minutes.

Evaluate

1) $37.37 + 13.73$

2) $61.1 - 46.46$

3) 0.375×4.8

4) $36.45 \div 0.9$

5) $9.5\times5.8 - 9.5\times4.8$

6) $33 \div 6.6 \times 5$

7) $96.96 - 68.68 \div 4$

8) $6^2 - 0.6^2 \div 0.3^2$

9) $0.4^2 \div (0.6 - 0.4)^2$

10) $12 \div [8.4 - (2.6^2 + 1.24)]$

Grade 8 Chapter 6 Lesson 1
Fractions and Mixed Numbers
Review 3 A

Date **Name**
Read directions carefully. **Answers must be in the indicated place.**
Show all steps for full credits. **Do only 1 operation per 1 step.**
Each problem is worth 5 points. **Solve 1 problem 5 minutes.**

1) Reduce the fraction $\frac{60}{100}$

Evaluate

2) $2\frac{5}{12}+1\frac{5}{6}$

3) $4-2\frac{3}{4}$

4) $2\frac{2}{5}\times 5\frac{5}{6}$

5) $4\frac{5}{8}\div 6\frac{1}{6}$

6) $\frac{2}{3}\div 1\frac{1}{3}\div\frac{5}{8}$

7) $\left(1\frac{1}{2}\right)^3\div\left(1\frac{1}{2}\right)$

8) $\left(\frac{1}{2}-\frac{1}{3}\right)^2\div\frac{1}{18}$

9) $6\frac{1}{2}\div\left(6\frac{1}{3}-4\frac{1}{6}\right)$

10) $1\frac{3}{4}\times 8-1\frac{1}{2}\times 8$

Grade 8 Chapter 6 Lesson 1
Converting Decimals and Fractions
Review 4 A

Date | Name
Read directions carefully. | Answers must be in the indicated place.
Show all steps for full credits. | Do only 1 operation per 1 step.
Each problem is worth 5 points. | Solve 1 problem 5 minutes.

1) Change terminating decimal 3.625 to fraction.

2) Change fraction $2\frac{3}{8}$ to terminating decimal.

Evaluate

3) $0.875 + \frac{3}{6}$

4) $1\frac{1}{4} - 0.45$

5) $1.6 \times \frac{3}{8}$

6) $0.15 \div \frac{3}{4}$

7) $3.75 \div 3\frac{1}{8}$

8) $3^3 \div \left(1\frac{1}{2}\right)^2$

9) $(0.3 - \frac{3}{20}) \div \frac{1}{20}$

10) $\frac{5}{6} \div (1\frac{2}{3} - 1.25)$

Grade 8 Chapter 6 Lesson 2
Rational Numbers
Review 5 A

Date | **Name**
Read directions carefully. | **Answers must be in the indicated place.**
Show all steps for full credits. | **Do only 1 operation per 1 step.**
Each problem is worth 5 points. | **Solve 1 problem 5 minutes.**

Evaluate

1) $\frac{1}{5}+(-2)$

2) $-(+1\frac{3}{10})-(-2\frac{7}{10})-(-\frac{1}{5})$

3) $(-1\frac{2}{5})(0.35)$

4) $(-4.5)\div(-1\frac{1}{2})$

5) $-(-\frac{1}{5})^2-(-\frac{2}{5})^2$

6) $-4\frac{4}{5}\times\frac{5}{8}-\frac{5}{8}\times3\frac{1}{5}$

7) $3\frac{1}{3}\div(-4\frac{5}{6}+3\frac{1}{2})$

8) $(\frac{1}{5}-0.3)\div0.2$

9) $0.15\div(-\frac{7}{10}+0.4)$

10) $\frac{3}{20}\div(-2.8+2\frac{1}{2})$

Grade 8 Chapter 6 Lesson 2
Equations
Review 6 A

Date **Name**

Read directions carefully. **Answers must be in the indicated place.**

Show all steps for full credits. **Do only 1 operation per 1 step.**

Each problem is worth 5 points. **Solve 1 problem 5 minutes.**

1) Simplify the expression $1.6(y - 4) - 0.6 - 2(0.3y - 7)$

2) Simplify the expression $4(\frac{1}{2}x - \frac{1}{4}y) - \frac{2}{3}(6x + 3y)$

Solve the equations

3) $\frac{1}{10}x + 10 = 50$

4) $81072 \div z = 9$

5) $28x - 987 - 45x = 9247$

6) $0.1y + 0.2y + 0.3y = 1.08$

7) $\frac{3}{10}y - \frac{1}{5}y - \frac{3}{10} = 1\frac{1}{5}$

8) $-5(\frac{2}{5}x + \frac{4}{5}) - 4(\frac{3}{4}x - \frac{1}{4}) = 0$

Solve the proportions

9) $\frac{7.8}{2.6} = \frac{4.5}{x}$

10) $5\frac{1}{4} \div 3\frac{1}{2} = 4\frac{1}{2} \div x$

Grade 8 Chapter 6 Lesson 2
Geometry
Review 7 A

Date | Name
Read directions carefully. | Answers must be in the indicated place.
Show all steps for full credits. | Do only 1 operation per 1 step.
Each problem is worth 5 points. | Solve 1 problem 5 minutes.

Solve the word problems

1) Find the perimeter of an isosceles triangle if its base is $3\frac{1}{2}$ in and leg $2\frac{1}{4}$ in.

2) Find one of leg of a right triangle if its other leg is 4.8 in and hypotenuse is 6 in.

3) Find the area of a square, if its side is $\frac{1}{2}$ in.

4) Find the area of a rectangle, if its length is $6\frac{2}{3}$ cm and the width is $4\frac{1}{2}$ cm.

5) Find the area of a trapezoid, if its height is $6\frac{2}{3}$ ft and bases are $9\frac{1}{2}$ ft and $8\frac{1}{2}$ ft.

6) Find the area of a triangle if its base is $6\frac{1}{4}$ m and height is $4\frac{4}{5}$ m.

7) Find the area of a right triangle if its legs are $4\frac{4}{5}$ ft and $5\frac{5}{6}$ ft.

8) Find the area of a triangle if its sides are 5 m, 12 m, and 13 m

Right Triangle $a^2 + b^2 = c^2$ $A = 0.5ab$

9) Find the circumference of a circle, if its diameter is 20 yd and $\pi \approx 3.14$

10) Find the volume of a cube if its side is $1\frac{1}{2}$ in.

Grade 8 Chapter 6 Lesson 2
Word Problems
Review 8 A

Date | **Name**
Read directions carefully. | **Answers must be in the indicated place.**
Show all steps for full credits. | **Do only 1 operation per 1 step.**
Each problem is worth 5 points. | **Solve 1 problem 5 minutes.**

Solve the word problems

1) A book has 80 pages. David read 0.35 of the whole book. How many pages more does he have left to read?

2) In the class, 7 students can dance. This equal $\frac{1}{4}$ of the whole class. How many students are in the class? How many students cannot dance?

3) Kristy read 30 pages. What part of the book did she read, if the book has 50 pages? Write the ratio as a reduced fraction. How many pages more does she have left to read?

4) Find the average for the numbers: $1\frac{1}{2}$, $1\frac{1}{5}$, $2\frac{1}{5}$, $2\frac{1}{10}$

5) A girl paid $24 for 4 books. How much would 6 of the same books cost?

6) A bus drove 65 mi at a speed of 25 mi/h. How much time it spent?

7) What speed does a boy have, if he drove 53 mi in 4 hours?

8) Of 32 children, 37.5% were boys. How many were girls?

9) A factory has 4 new machines and 80% were old machine. How many machines are in the factory?

10) In 50 minutes of lesson, students solve an equation in 25 minutes. What percent of lesson time did students use to solve the equation?

Grade 8 Chapter 6 Lesson 3
Whole Numbers
Review 1 B

Date **Name**

Read directions carefully. **Answers must be in the indicated place.**

Show all steps for full credits. **Do only 1 operation per 1 step.**

Each problem is worth 5 points. **Solve 1 problem 5 minutes.**

Evaluate

1) $(100 - 20) \times 25$

2) $14528 \div 16 - 9680 \div 16$

3) $58800 \div (95^2 - 25^2)$

4) 2305×908

5) $380304 \div 76$

6) $6^3 - 6^2 \div 2^2$

7) $8^2 \div 4^2 - 2^2$

8) $3^4 \div (6 - 3)^2$

9) $6^3 - 2^3(10 - 2^3 \div 2^2)$

10) $\{3 \cdot 10^2 - 2^2[2^2 + (2 + 3)^2]\} + 2^3$

Grade 8 Chapter 6 Lesson 3
Decimals
Review 2 B

Date **Name**

Read directions carefully. **Answers must be in the indicated place.**
Show all steps for full credits. **Do only 1 operation per 1 step.**
Each problem is worth 5 points. **Solve 1 problem 5 minutes.**

Evaluate

1) $46.46 + 14.64$

2) $51.1 - 37.37$

3) 0.375×2.4

4) $45.54 \div 0.9$

5) $8.6\times4.7 - 8.6\times3.7$

6) $33 \div 5.5 \times 6$

7) $48.48 - 34.34 \div 2$

8) $4^2 - 0.4^2 \div 0.2^2$

9) $0.6^2 \div (0.5 - 0.3)^2$

10) $9.6 \div [\ 11.4 - (2.8^2 + 1.16)]$

Grade 8 Chapter 6 Lesson 3
Fractions and Mixed Numbers
Review 3 B

Date | **Name**

Read directions carefully.
Show all steps for full credits.
Each problem is worth 5 points.

Answers must be in the indicated place.
Do only 1 operation per 1 step.
Solve 1 problem 5 minutes.

Evaluate

1) Reduce the fraction $\frac{40}{100}$

2) $2\frac{1}{12}+1\frac{1}{6}$

3) $4-1\frac{1}{5}$

4) $1\frac{5}{6}\times 2\frac{2}{11}$

5) $3\frac{3}{8}\div 5\frac{5}{8}$

6) $\frac{1}{6}\div 1\frac{1}{3}\div\frac{5}{8}$

7) $(1\frac{1}{2})^2\div\left(1\frac{1}{4}\right)$

8) $\left(\frac{1}{3}-\frac{1}{6}\right)^2\div\frac{1}{9}$

9) $4\frac{1}{2}\div(6\frac{1}{2}-4\frac{1}{4})$

10) $1\frac{1}{2}\times 6-1\frac{1}{6}\times 6$

Grade 8 Chapter 6 Lesson 3
Converting Decimals and Fractions
Review 4 B

Date **Name**
Read directions carefully. **Answers must be in the indicated place.**
Show all steps for full credits. **Do only 1 operation per 1 step.**
Each problem is worth 5 points. **Solve 1 problem 5 minutes.**

Evaluate

1) **Change terminating decimal 2.375 to fraction.**

2) **Change fraction $3\frac{5}{8}$ to terminating decimal.**

Evaluate

3) $1\frac{3}{6}+1.125$

4) $2\frac{1}{4}-1.75$

5) $3.2\times\frac{5}{8}$

6) $0.45\div\frac{3}{4}$

7) $4\frac{1}{8}\div 2.75$

8) $(1\frac{1}{2})^2\div 3^2$

9) $(2.5-2\frac{1}{3})\div 1\frac{1}{3}$

10) $6.5\div(4\frac{1}{3}-2\frac{1}{6})$

Grade 8 Chapter 6 Lesson 4
Rational Numbers
Review 5 B

Date **Name**
Read directions carefully. **Answers must be in the indicated place.**
Show all steps for full credits. **Do only 1 operation per 1 step.**
Each problem is worth 5 points. **Solve 1 problem 5 minutes.**

Evaluate

1) $-4-(-3\frac{2}{5})$

2) $-\frac{1}{5}-(-\frac{3}{10})-(-2\frac{1}{10})$

3) $-1\frac{4}{5}(0.45)$

4) $1.2\div(-1\frac{1}{2})$

5) $-(-\frac{3}{5})^2-(-\frac{4}{5})^2$

6) $-3\frac{2}{3}\times1\frac{1}{2}+1\frac{2}{3}\times1\frac{1}{2}$

7) $1\frac{11}{16}\div(-4\frac{1}{4}+3\frac{1}{8})$

8) $(-3\frac{1}{5}+1.1)\div4.2$

9) $\frac{8}{25}\div(-0.8+\frac{2}{5})$

10) $\frac{7}{20}\div(-0.3+\frac{1}{20})$

Grade 8 Chapter 6 Lesson 4
Equations
Review 6 B

Date **Name**

Read directions carefully. **Answers must be in the indicated place.**

Show all steps for full credits. **Do only 1 operation per 1 step.**

Each problem is worth 5 points. **Solve 1 problem 5 minutes.**

1) Simplify the expression $3z + 2(2z - 3) - 8 - 7(z - 2)$

2) Simplify the expression $\frac{4}{9}(18x - 9y) - 12(\frac{5}{6}x - \frac{1}{6}y)$

Solve equations

3) $\frac{1}{11}x - 5 = 55$

4) $72056 \div z = 8$

5) $125x - 876 - 201x = 37580$

6) $0.1z + 0.2z + 0.3z = 1.5$

7) $\frac{4}{5}z - 0.3z - 3 = 1.5$

8) $1.6(\frac{3}{4}y - \frac{5}{8}) - 1.8(\frac{2}{9}y - \frac{7}{9}) = 0$

Solve the proportions

9) $\frac{4.4}{z} = \frac{8.8}{2.2}$

10) $3\frac{1}{3} \div 2\frac{1}{2} = 2\frac{2}{3} \div x$

Grade 8 Chapter 6 Lesson 4
Geometry
Review 7 B

Date | **Name**
Read directions carefully. | **Answers must be in the indicated place.**
Show all steps for full credits. | **Do only 1 operation per 1 step.**
Each problem is worth 5 points. | **Solve 1 problem 5 minutes.**

Solve the geometry problems

1) Find the perimeter of an isosceles triangle if its base is $4\frac{1}{2}$ in and leg is $3\frac{1}{4}$ in.

2) Find the leg of right triangle if its hypotenuse is 9.5 m and other leg is 5.7 m.

3) Find the area of a square if its side is $3\frac{1}{2}$ in.

4) Find the area of a parallelogram if its base is $6\frac{1}{4}$ cm and height is $4\frac{4}{5}$ cm.

5) Find the area of a trapezoid, if its height is $4\frac{2}{3}$ ft and bases are $8\frac{1}{2}$ ft and $6\frac{1}{2}$ ft.

6) Find the area of a triangle if its base is $8\frac{1}{3}$ in and height is $7\frac{1}{5}$ in.

7) Find the area of a right triangle if its legs are $4\frac{1}{5}$ ft and $6\frac{2}{3}$ ft.

8) Find the area of a triangle if its sides are 15 m, 20 m, and 25 m

Right Triangle $a^2 + b^2 = c^2$ $A = 0.5ab$

9) Find the circumference of a circle, if its diameter is 10 yd and $\pi \approx 3.14$

10) Find the volume of a cube if its side is $2\frac{1}{2}$ yd.

Grade 8 Chapter 6 Lesson 4
Word Problems
Review 8 B

Date **Name**

Show all steps for full credits. Each problem is worth 5 points.
Do only 1 operation per 1 step. Solve 1 problem 5 minutes.

Solve the word problems

1) A book has 60 pages. David read 0.75 of the whole book. How many pages more does he have left to read?

2) In the class, 12 students can dance; this is $\frac{2}{5}$ of the whole class. How many students are in the class? Write the ratio as a reduced fraction. How many students cannot dance?

3) Kristy read 24 pages. What part of the book did she read, if the book has 40 pages? Write the ratio as a reduced fraction. How many pages more does she have left to read?

4) Find the average for the numbers: $1\frac{1}{2}$, $1\frac{1}{5}$, $4\frac{1}{5}$, $4\frac{1}{10}$

5) A girl paid $12 for 2 books. How much would 3 of the same books cost?

6) A bus drove 55 mi at a speed of 25 mi/h. How much time it spent?

7) What speed does a boy have, if he drove 69 mi in 4 hours?

8) Of 16 children, 37.5% were boys. How many were girls?

9) A factory has 8 new machines and 60% were old machine. How many machines are in the factory?

10) In 75 minutes of lesson, students solve an equation in 45 minutes. What percent of lesson time did students use to solve the equation?

Grade 8 Chapter 6 Lesson 5
Whole Numbers
Review 1 C

Date — **Name**

Read directions carefully. **Answers must be in the indicated place.**
Show all steps for full credits. **Do only 1 operation per 1 step.**
Each problem is worth 5 points. **Solve 1 problem 5 minutes.**

Evaluate

1) $\begin{array}{r} 340506 \\ +\ 405067 \\ \hline \end{array}$

2) $\begin{array}{r} 800000 \\ -\ 776655 \\ \hline \end{array}$

3) $\begin{array}{r} 807 \\ \times 504 \\ \hline \end{array}$

4) $\begin{array}{r} 3405 \\ \times\ 708 \\ \hline \end{array}$

5) $522435 \div 87$

6) $488235 \div 807$

7) $226240 \div 70 \times 8$

8) $6400 - 5648 \div 8$

9) $300 - 240 \div 6 \times 5$

10) $90 \div 3 \times 2 - 60 \div 4 \times 3$

Grade 8 Chapter 6 Lesson 5
Decimals
Review 2 C

Date **Name**
Read directions carefully. **Answers must be in the indicated place.**
Show all steps for full credits. **Do only 1 operation per 1 step.**
Each problem is worth 5 points. **Solve 1 problem 5 minutes.**

Evaluate

1) $19.19 + 11.91$

2) $41.1 - 28.28$

3) 0.625×1.6

4) $18.36 \div 0.6$

5) $7.8\times6.2 - 7.8\times5.2$

6) $22 \div 5.5 \times 4$

7) $48.48 - 24.24 \div 2$

8) $3^2 - 0.9^2 \div 0.3^2$

9) $0.2^2 \div (0.4 - 0.2)^2$

10) $44 \div [\, 6.2 - (1.6^2 + 1.44)]$

Grade 8 Chapter 6 Lesson 5
Fractions and Mixed Numbers
Review 3 C

Date | Name
Read directions carefully. | Answers must be in the indicated place.
Show all steps for full credits. | Do only 1 operation per 1 step.
Each problem is worth 5 points. | Solve 1 problem 5 minutes.

1) Reduce the fraction $\frac{50}{100}$

Evaluate

2) $1\frac{6}{15}+2\frac{1}{5}$

3) $5-1\frac{1}{5}$

4) $3\frac{3}{4}\times 5\frac{1}{3}$

5) $2\frac{7}{8}\div 3\frac{5}{6}$

6) $1\frac{1}{4}\div\frac{3}{4}\div 1\frac{2}{3}$

7) $\left(1\frac{1}{2}\right)^4\div\left(1\frac{1}{2}\right)^2$

8) $\left(\frac{1}{4}-\frac{1}{8}\right)^2\div\frac{1}{32}$

9) $2\frac{1}{2}\div\left(3\frac{1}{2}-2\frac{1}{4}\right)$

10) $2\frac{1}{2}\times 16-1\frac{1}{4}\times 16$

Grade 8 Chapter 6 Lesson 5
Converting Decimals and Fractions
Review 4 C

Date **Name**
Read directions carefully. **Answers must be in the indicated place.**
Show all steps for full credits. **Do only 1 operation per 1 step.**
Each problem is worth 5 points. **Solve 1 problem 5 minutes.**

1) Change decimal 1.625 to fraction.

2) Change fraction $1\frac{7}{8}$ to terminating decimal.

Evaluate

3) $0.25 + \frac{4}{8}$

4) $1\frac{1}{4} - 0.75$

5) $2.4 \times \frac{3}{8}$

6) $0.36 \div \frac{3}{4}$

7) $4.25 \div 2\frac{1}{8}$

8) $3^2 \div \left(1\frac{1}{2}\right)^2$

9) $(0.6 - \frac{3}{20}) \div \frac{3}{20}$

10) $\frac{5}{6} \div (2\frac{2}{3} - 2.25)$

Grade 8 Chapter 6 Lesson 6
Rational Numbers
Review 5 C

Date | Name

Read directions carefully. | Answers must be in the indicated place.

Show all steps for full credits. | Do only 1 operation per 1 step.

Each problem is worth 5 points. | Solve 1 problem 5 minutes.

Evaluate

1) $\frac{2}{5}+(-3)$

2) $-(+2\frac{7}{10})-(-1\frac{3}{10})-(-\frac{3}{5})$

3) $(-1\frac{3}{5})(0.75)$

4) $(4.5)\div(-1\frac{1}{4})$

5) $(-\frac{2}{5})^2-(-\frac{3}{5})^2$

6) $-\frac{5}{6}\times4\frac{1}{5}+\frac{5}{6}\times1\frac{1}{5}$

7) $4\frac{1}{2}\div(-4\frac{5}{6}+3\frac{1}{3})$

8) $(\frac{1}{5}-0.4)\div0.5$

9) $0.25\div(-\frac{8}{10}+0.3)$

10) $5\frac{1}{5}\div(-2.6+2\frac{1}{2})$

Grade 8 Chapter 6 Lesson 6
Equations
Review 6 C

Date **Name**

Read directions carefully. **Answers must be in the indicated place.**
Show all steps for full credits. **Do only 1 operation per 1 step.**
Each problem is worth 5 points. **Solve 1 problem 5 minutes.**

1) **Simplify the expression** $10\,(0.1x - 4) - 100\,(0.01x - 5)$

2) **Simplify the expression** $16(\frac{3}{4}x + \frac{1}{4}y) - \frac{2}{3}(9x + 27y)$

Solve the equations

3) $\frac{1}{8}x - 10 = 40$

4) $49056 \div z = 7$

5) $-78x - 678 + 9x = 20298$

6) $0.8z - 0.3z - 3 = 1.5$

7) $\frac{1}{10}y + 5 + \frac{1}{10}y = 10$

8) $-6(\frac{1}{6} - \frac{1}{2}x) - 2(1 - \frac{1}{2}x) = 0$

Solve the proportions

9) $\frac{y}{5.1} = \frac{2.7}{0.9}$

10) $z \div \frac{7}{12} = \frac{2}{3} \div \frac{7}{9}$

Grade 8 Chapter 6 Lesson 6
Geometry
Review 7 C

Date | **Name**
Read directions carefully. | **Answers must be in the indicated place.**
Show all steps for full credits. | **Do only 1 operation per 1 step.**
Each problem is worth 5 points. | **Solve 1 problem 5 minutes.**

Solve the geometry problems

1) Find the perimeter of a isosceles triangle if its base $6\frac{1}{2}$ in and leg $4\frac{1}{4}$ in.

2) Find the hypotenuse of a right triangle if its legs are 5.2 m and 3.9 m.

3) Find the area of a square, if its side is 2.5 in.

4) Find the area of a rectangle, if its length is $10\frac{1}{2}$ cm and width is $5\frac{1}{3}$ cm.

5) Find the area of a trapezoid, if its height is $5\frac{1}{3}$ ft, bases are $6\frac{1}{2}$ ft and $5\frac{1}{2}$ ft.

6) Find the area of a triangle if its base is $9\frac{3}{4}$ in and height is $5\frac{1}{3}$ in.

7) Find the area of a right triangle if its legs are $4\frac{4}{5}$ ft and $6\frac{1}{4}$ ft.

8) Find the area of a triangle if its sides are 6 m, 8 m, 10 m.

Right Triangle $a^2 + b^2 = c^2$ $A = 0.5ab$

9) Find the circumference of a circle, if its radius is 10 ft, $\pi \approx 3.14$.

10) Find exact answer of volume of a sphere with radius is 3 yd.

Grade 8 Chapter 6 Lesson 6
Word Problems
Review 8 C

Date	**Name**
Read directions carefully.	**Answers must be in the indicated place.**
Show all steps for full credits.	**Do only 1 operation per 1 step.**
Each problem is worth 5 points.	**Solve 1 problem 5 minutes.**

Solve the word problems

1) A book has 80 pages. David read 0.45 of the whole book. How many pages more does he have left to read?

2) In the class, 8 students can dance; this is $\frac{1}{5}$ of the whole class. How many students are in the class? How many students cannot dance?

3) Kristy read 40 pages. What part of the book did she read, if the book has 50 pages? Write the ratio as a reduced fraction. How many pages more does she have left to read?

4) Find the average for the numbers: $2\frac{1}{2}$, $2\frac{1}{5}$, $1\frac{1}{5}$, $1\frac{1}{10}$

5) A girl paid $15 for 6 books. How much would 8 of the same books cost?

6) A bus drove 75 mi at a speed of 30 mi/h. How much time it spent?

7) What speed does a boy have, if he drove 57 mi in 4 hours?

8) Of 32 children, 62.5% were boys. How many were girls?

9) A factory has 5 new machines and 80% were old machine. How many machines are in the factory?

10) In 50 minutes of lesson, students solve an equation in 30 minutes. What percent of lesson time did students use to solve the equation?

Grade 8 Chapter 6 Lesson 7
Whole Numbers
Review 1 D

Date **Name**

Read directions carefully. **Answers must be in the indicated place.**
Show all steps for full credits. **Do only 1 operation per 1 step.**
Each problem is worth 5 points. **Solve 1 problem 5 minutes.**

Evaluate

1) $(200 - 140) \times 25$

2) $12912 \div 16 - 11296 \div 16$

3) $39600 \div (85^2 - 25^2)$

4) 2406×908

5) $260195 \div 65$

6) $4^3 - 4^2 \div 2^2$

7) $6^3 \div 2^3 - 3^3$

8) $3^3 \div (6 - 3)^2$

9) $7^3 - 2^2(3^2 \times 2^3 - 2^2)$

10) $\{3 \cdot 8^2 - 2^2[2^3 + (1 + 3)^2]\} + 2^3$

Grade 8 Chapter 6 Lesson 7

Decimals

Review 2 D

Date | **Name**

Read directions carefully. **Answers must be in the indicated place.**

Show all steps for full credits. **Do only 1 operation per 1 step.**

Each problem is worth 5 points. **Solve 1 problem 5 minutes.**

Evaluate

1) $28.28 + 12.82$

2) $31.1 - 19.19$

3) 0.625×3.2

4) $24.36 \div 0.6$

5) $6.8\times3.9 - 6.8\times2.9$

6) $22 \div 4.4 \times 5$

7) $52.52 - 36.36 \div 4$

8) $2^2 - 0.4^2 \div 0.2^2$

9) $0.3^2 \div (0.6 - 0.3)^2$

10) $48 \div [8.4 - (1.8^2 + 2.76)]$

Grade 8 Chapter 6 Lesson 7
Fractions and Mixed Numbers
Review 3 D

Date | **Name**
Read directions carefully. | **Answers must be in the indicated place.**
Show all steps for full credits. | **Do only 1 operation per 1 step.**
Each problem is worth 5 points. | **Solve 1 problem 5 minutes.**

1) Reduce the fraction $\frac{20}{100}$

Evaluate

2) $1\frac{6}{15}+1\frac{2}{5}$

3) $3-1\frac{1}{4}$

4) $4\frac{2}{5}\times 1\frac{4}{11}$

5) $3\frac{3}{4}\div 6\frac{1}{4}$

6) $\frac{3}{8}\div\frac{3}{4}\div 1\frac{1}{4}$

7) $(1\frac{4}{5})\div\left(1\frac{1}{2}\right)^2$

8) $\left(\frac{1}{5}-\frac{1}{10}\right)^2\div\frac{1}{10}$

9) $1\frac{1}{2}\div(5\frac{1}{2}-4\frac{1}{4})$

10) $1\frac{3}{4}\times 12-1\frac{1}{4}\times 12$

Grade 8 Chapter 6 Lesson 7
Converting Decimals and Fractions
Review 4 D

Date **Name**
Read directions carefully. **Answers must be in the indicated place.**
Show all steps for full credits. **Do only 1 operation per 1 step.**
Each problem is worth 5 points. **Solve 1 problem 5 minutes.**

1) Change terminating decimal 2.875 to fraction.

2) Change fraction $2\frac{5}{8}$ to terminating decimal.

Evaluate

3) $1\frac{4}{8}+1.25$

4) $2\frac{1}{4}-1.75$

5) $4.8\times\frac{5}{8}$

6) $0.57\div\frac{3}{4}$

7) $6\frac{7}{8}\div 2.75$

8) $3^4\div(1\frac{1}{2})^2$

9) $(3.5-3\frac{1}{3})\div 1\frac{1}{3}$

10) $6.5\div(3\frac{1}{3}-1\frac{1}{6})$

Grade 8 Chapter 6 Lesson 8
Rational Numbers
Review 5 D

Date **Name**
Read directions carefully. **Answers must be in the indicated place.**
Show all steps for full credits. **Do only 1 operation per 1 step.**
Each problem is worth 5 points. **Solve 1 problem 5 minutes.**

Evaluate

1) $-5-(-4\frac{2}{5})$

2) $-\frac{2}{5}-(-\frac{7}{10})-(-2\frac{3}{10})$

3) $-1\frac{1}{5}(0.25)$

4) $1.5\div(-1\frac{1}{4})$

5) $-(-\frac{4}{5})^2+(-\frac{1}{5})^2$

6) $-3\frac{1}{3}\times1\frac{1}{5}-1\frac{1}{5}\times1\frac{2}{3}$

7) $1\frac{1}{4}\div(-4\frac{5}{8}+3\frac{1}{16})$

8) $(-2\frac{1}{5}+0.6)\div3.2$

9) $-\frac{4}{25}\div(-0.6+\frac{4}{5})$

10) $-\frac{9}{20}\div(-0.3+\frac{7}{20})$

Grade 8 Chapter 6 Lesson 8
Equations
Review 6 D

Date **Name**

Read directions carefully. **Answers must be in the indicated place.**

Show all steps for full credits. **Do only 1 operation per 1 step.**

Each problem is worth 5 points. **Solve 1 problem 5 minutes.**

1) **Simplify the expression** $100\ (0.01y - 0.03) - 1000\ (0.001y - 0.05)$

2) **Simplify the expression** $\frac{3}{7}(14x - 21y) - 15(\frac{2}{3}x - \frac{1}{3}y)$

Solve the equations

3) $\frac{1}{9}x + 15 = 45$

4) $64072 \div z = 8$

5) $9y + 789 - 28y = 8541$

6) $0.6x - 0.3x - 0.2 = 2.5$

7) $0.1z - \frac{2}{5}z + \frac{1}{2}z = 1.5$

8) $-8(\frac{1}{2} - \frac{1}{4}y) - 3(\frac{2}{3} - \frac{1}{3}y) = 0$

Solve the proportions

9) $\frac{y}{4.7} = \frac{27.3}{9.1}$

10) $2 \div x = 1\frac{4}{7} \div 3\frac{1}{7}$

Grade 8 Chapter 6 Lesson 8
Geometry
Review 7 D

Date | **Name**
Read directions carefully. | **Answers must be in the indicated place.**
Show all steps for full credits. | **Do only 1 operation per 1 step.**
Each problem is worth 5 points. | **Solve 1 problem 5 minutes.**

Solve the geometry problems

1)Find the perimeter of an isosceles triangle if its base $7\frac{1}{2}$ in and leg is $5\frac{1}{4}$ in.

2) Find the hypotenuse of a right triangle if its legs are 4.2 m and 5.6 m.

3) Find the area of a square, if its side is 1.5 in.

4) Find the area of a rectangle, if its length is $7\frac{1}{2}$ cm, width is $4\frac{2}{3}$ cm.

5) Find the area of trapezoid, if its height is $2\frac{2}{3}$ ft, bases are $7\frac{1}{2}$ ft and $4\frac{1}{2}$ ft.

6) Find the area of a triangle if its base is $8\frac{3}{4}$ m and height is $6\frac{2}{5}$ m.

7) Find the area of a right triangle if its legs are $5\frac{2}{5}$ ft and $6\frac{2}{3}$ ft.

8) Find the area of a triangle if its sides are 9 in, 12 in, and 15 in.

Right Triangle $a^2 + b^2 = c^2$ $A = 0.5ab$

9) Find the circumference of a circle, if its radius is 20 ft, $\pi \approx 3.14$

10) Find exact answer of volume of a sphere with radius is 6 yd.

Grade 8 Chapter 6 Lesson 8
Word Problems
Review 8 D

Date **Name**
Read directions carefully. **Answers must be in the indicated place.**
Show all steps for full credits. **Do only 1 operation per 1 step.**
Each problem is worth 5 points. **Solve 1 problem 5 minutes.**

Solve the word problems

1) A book has 60 pages. David read 0.85 of the whole book. How many pages more does he have left to read?

2) In the class, 15 students can dance; this is $\frac{3}{5}$ of the whole class. How many students are in the class? How many students cannot dance?

3) Kristy read 32 pages. What part of the book did she read, if the book has 40 pages? Write the ratio as a reduced fraction. How many pages more does she have left to read?

4) Find the average for the numbers: $4\frac{1}{2}$, $4\frac{1}{5}$, $1\frac{1}{5}$, $1\frac{1}{10}$.

5) A girl paid $12 for 3 books. How much would 4 of the same books cost?

6) A bus drove 63 mi at a speed of 30 mi/h. How much time it spent?

7) What speed does a boy have, if he drove 73 mi in 4 hours?

8) Of 16 children, 62.5% were boys. How many were girls?

9) A factory has 10 new machines and 60% were old machine. How many machines are in the factory?

10) In 75 minutes of lesson, students solve an equation in 60 minutes. What percent of lesson time did students use to solve the equation?

College Math Chapter 6 Lesson 9
Worksheet 9
Many digit numbers

Date **Name**

Read directions carefully. **Answers must be in the indicated place.**
Show all steps for full credits. **Do only 1 operation per 1 step.**
Each problem is worth 5 points. **Solve 1 problem 5 minutes.**

Evaluate

1) $$\begin{array}{r} 700000 \\ \underline{-\ 234567} \end{array}$$

2) $$\begin{array}{r} 765 \\ \underline{\times 604} \end{array}$$

3) $$\begin{array}{r} 706 \\ \underline{\times 605} \end{array}$$

4) $193884 \div 321$

5) $243714 \div 302$

6) $240240 \div 80 \times 3$

7) $7490 - 830 \times 3$

8) $56049 - 49042 \div 7$

9) $(30 \times 560) \div 8$

10) $289 \div (93 - 76)$

College Math Chapter 6 Lesson 10
Worksheet 10
Many digit numbers

Date | Name
Read directions carefully. | Answers must be in the indicated place.
Show all steps for full credits. | Do only 1 operation per 1 step.
Each problem is worth 5 points. | Solve 1 problem 5 minutes.

Evaluate

1) $\begin{array}{r} 600000 \\ \underline{-\ 123456} \end{array}$

2) $\begin{array}{r} 654 \\ \underline{\times 503} \end{array}$

3) $\begin{array}{r} 605 \\ \underline{\times 403} \end{array}$

4) $105630 \div 210$

5) $182508 \div 201$

6) $180180 \div 90 \times 2$

7) $6840 - 920 \times 2$

8) $42036 - 36030 \div 6$

9) $(20 \times 720) \div 9$

10) $256 \div (81 - 65)$

Grade 8 Answers Chapter 6

Memorize Exponents

1)1 2) 4 3) 9 4) 16 5) 25 6) 36 7) 49 8) 64 9) 81 10) 100
11)1 12) 8 13) 27 14) 64 15) 125 16) 216 17) 343 18) 512 19) 729 20) 1000
21) 2 22) 4 23) 8 24) 16 25) 32 26) 64 27) 128 28) 256 29) 512 30) 1024

Lesson 1 College Placement Test

1) 5 2) 0 3) 21 4) - 2.8 5) - 4x – 4y
6) 7.5 7) 18.2 ft 8) 60 in^2 9) 2.4 hr. 10) 14 girl

Lesson 1 Review 1 A

1) 956,6852) 12,234 3) 549,340 4) 3,645,354 5) 7,006
6) 706 7) 40,905 8) 7,293 9) 300 10) 0

Lesson 1 Review 2 A

1) 51.1 2) 14.64 3) 1.8 4) 40.5 5) 9.5
6) 25 7) 79.79 8) 32 9) 4 10) 30

Lesson 1 Review 3 A

1) $\frac{3}{5}$ 2) 4.25 3) 1.25 4) 14 5) 0.75
6) 0.8 7) 2.25 8) 0.5 9) 3 10) 2

Lesson 1 Review 4 A

1) $3\frac{5}{8}$ 2) 2.375 3) 1.375 4) 0.8 5) 0.6
6) 0.2 7) 1.2 8) 12 9) 3 10) 2

Lesson 2 Review 5 A

1) -1.8 2) 1.6 3) -0.49 4) -3 5) - 0.2
6) - 5 7) - 2.5 8) - 0.5 9) - 0.5 10) - 0.5

Lesson 2 Review 6 A

1) y + 7 2) - 2x – 3y 3) 400 4) 9008 5) - 602
6) 1.8 7) 15 8) - 0.6 9) 1.5 10) 3

Lesson 2 Review 7 A

1) 8 in 2) 3.6 in 3) 0.25in^2 4) 30 cm^2 5) 60 ft^2
6) 15 $m^2$7) 14 ft^2 8) 30 m^2 9) 62.8 yd 10) 3.375 in^3

Lesson 2 Review 8 A

1) 52 pages 2) 28 students, 21 students 3) $\frac{3}{5}$, 20 pages 4) 1.75
5) $36 6) 2.6 hr 7) 13.25 mi/hr. 8) 20 girls
9) 20 machines 10) 50%

Grade 8 Answers Chapter 6

Lesson 3 Review 1 B

1) 2000	2) 303	3) 7	4) 2092940	5) 5004
6) 207	7) 0	8) 9	9) 152	10) 192

Lesson 3 Review 2 B

1) 61.1	2) 13.73	3) 0.9	4) 50.6	5) 8.6
6) 36	7) 31.31	8) 12	9) 9	10) 4

Lesson 3 Review 3 B

1) $\frac{2}{5}$	2) 3.25	3) 2.8	4) 4	5) 0.6
6) 0.2	7) 1.8	8) 0.25	9) 2	10) 2

Lesson 3 Review 4 B

1) $2\frac{3}{8}$	2) 3.625	3) 2.625	4) 0.5	5) 2
6) 0.6	7) 1.5	8) 0.25	9) 0.125	10)3

Lesson 4 Review 5 B

1) -0.6	2) 2.2	3) - 0.81	4) - 0.8	5) - 1
6) - 3	7) - 1.5	8) - 0.5	9) - 0.8	10) - 1.4

Lesson 4 Review 6 B

1) 0	2) - 2x – 2y	3) 660	4) 9007	5) - 506
6) 2.5	7) 9	8) - 0.5	9) 1.1	10) 2

Lesson 4 Review 7 B

1) 11 in	2) 7.6 m	3) 12.25 in^2	4) 30 cm^2	5) 35 ft^2
6) 30 in^2	7) 14 ft^2	8) 150 m^2	9) 31.4 yd	10) 15.625 yd^3

Lesson 4 Review 8 B

1) 15 pages 2) 30 students, 18 students 3) $\frac{3}{5}$, 16 pages

4) 2.75 5) $ 18 6) 2.2 hr 7) 17.25 mi/hr 8) 10 girls

9) 20 machines 10) 60%

Grade 8 Answers Chapter 6

Lesson 5 Review 1 C

1) 745,573 2) 23,345 3) 406,728 4) 2,410,740 5) 6,005
6) 605 7) 25,856 8) 5,694 9) 100 10) 15

Lesson 5 Review 2 C

1) 31.1 2) 12.82 3) 1 4) 30.6 5) 7.8
6) 16 7) 36.36 8) 0 9) 1 10) 20

Lesson 5 Review 3 C

1) $\frac{1}{2}$ 2) 3.6 3) 3.8 4) 20 5) 0.75
6) 1 7) 2.25 8) 0.5 9) 2 10) 20

Lesson 5 Review 4 C

1) $1\frac{5}{8}$ 2) 1.875 3) 0.75 4) 0.5 5) 0.9
6) 0.48 7) 2 8) 4 9) 3 10) 2

Lesson 6 Review 5 C

1) -2.6 2) -0.8 3) - 1.2 4) - 3.6 5) -0.2
6) -2.5 7) - 3 8) -0.4 9) -0.5 10) -52

Lesson 6 Review 6 C

1) 460 2) 6x – 14y 3) 400 4) 7008 5) - 304
6) 9 7) 25 8) 0.75 9) 15.3 10) 0.5

Lesson 6 Review 7 C

1) 15 in 2) 6.5 m 3) 6.25 in^2 4) 56 cm^2 5) 32 ft^2
6) 26 in^2 7) 15 ft^2 8) 24 m^2 9) 62.8 ft 10) 36π yd^3

Lesson 6 Review 8 C

1) 44 pages 2) 40 students, 32 students 3) $\frac{4}{5}$, 10 pages 4) 1.75 5) $ 20
6) 2.5 hr 7) 14.25 mi/hr 8) 12 girls 9) 25 machines 10) 60%

Grade 8 Answers Chapter 6

Lesson 7 Review 1 D

1) 1500 2) 101 3) 6 4) 2,184,648 5) 4003
6) 60 7) 0 8) 3 9) 71 10) 104

Lesson 7 Review 2 D

1) 41.1 2) 11.91 3) 2 4) 40.6 5) 6.8
6) 25 7) 43.43 8) 0 9) 1 10) 20

Lesson 7 Review 3 D

1) $\frac{1}{5}$ 2) 2.8 3) 1.75 4) 6 5) 0.6
6) 0.4 7) 0.8 8) 0.1 9) 1.2 10) 6

Lesson 7 Review 4 D

1) $2\frac{7}{8}$ 2) 2.625 3) 2.75 4) 0.5 5) 3
6) 0.76 7) 2.5 8) 36 9) 0.125 10) 3

Lesson 8 Review 5 D

1) - 0.6 2) 2.6 3) -0.3 4) -1.2 5) -0.6
6) -6 7) -0.8 8) -0.5 9) -0.8 10) -9

Lesson 8 Review 6 D

1) 47 2) $-4x-4y$ 3) 270 4) 8009 5) - 408
6) 9 7) 7.5 8) 2 9) 14.1 10) 4

Lesson 8 Review 7 D

1) 18 in 2) 7 m 3) 2.25 in^2 4) 35 cm^2 5) 16 ft^2
6) 28 m^2 7) 18 ft^2 8) 54 in^2 9) 125.6 ft 10) 288π yd^3

Lesson 8 Review 8 D

1) 9 pages 2) 10 students 3) 0.8, 8 pages 4) 2.75 5) $ 16
6) 2.1 hr. 7) 18.25 mi/hr. 8) 6 girls 9) 25 machines 10) 80%

Lesson 9

Worksheet 9

1) 465433 2) 462060 3) 427130 4) 604 5) 807
6) 9009 7) 5000 8) 49043 9) 2100 10) 17

Lesson 10

Worksheet 10

1) 476544 2) 328962 3) 243815 4) 503 5) 908
6) 4004 7) 5000 8) 36031 9) 1600 10) 16

Students!! Attention!!

Adults, help please students 1 – 4 grades to read and understand advice.

When I was in 5th grade I did not like math. I am not always understood math. I was 11 years girl. Next 6th grade I got a very good teacher. She explains very clearly and I understood her very well. I was worked very hard and I was not afraid to ask questions. Final Grade in Algebra in 6th grade I got an A. I said in class that I want to be a teacher of math. But some students said that I am not enough smart to be a math teacher. I said I will be work hard with math. I did it, I got a Master's degree in math and education, I love my students, I was teaching many years, in High School, part-time college math instructor, full-time university math lecturer.

Some people do a simple job, eat, sleep, go to work. People with a bad brain have not so a good job, not so good salary. If you look at TV how someone's dance, you will be not changing your brain. But if you take a dancing class this helps your body and brain ability. The same if you want to use the calculator, but the calculator not always with you. Much better if you learn how you need to do any operations with numbers without a calculator. If you read this math book you will know what you need to do. This teaches you how to improve your memory, attention, logic thinking, how to read your math books every day, your brain works better. You can work 50 minutes with math, 10 minutes break maybe dance. Do not afraid to do mistakes, fixed and move to the next step. Don't listen to anyone, choose what do you like more.

Students need to know if you work with the brain every day, your brain works better. You need a good brain in any job, in any class. When you finish High School, you need to get any profession, you need to go to college and get a good job and a better salary. You need to learn any subject, its work for the brain. Read the book, do the exercises yourself, try your best, and then check your answer at the end of the chapter. Be honest with yourself. You need to believe in yourself. The brain is like muscles. People with good brain lives longer. The professors at my university were over 80 years old, wise, and optimistic. It was a great pleasure to talk to them.